Traffic Monitoring: Automobiles, Trucks, Bicycles, and Pedestrians 2018

Transportation Research Record: Journal of the Transportation Research Board Volume 2672, Issue 43

ISBN: 978-1-5108-8249-2

Printed from e-media with permission by:

Curran Associates, Inc.
57 Morehouse Lane
Red Hook, NY 12571

Some format issues inherent in the e-media version may also appear in this print version.

Published by Sage Publications USA

ISBN (Print) 978-1-5108-8249-2
ISBN 2018 Printed Set (All Issues) 978-1-5108-7735-1

Printed by Curran Associates, Inc. (2019)

Additional copies of this publication are available from:

Curran Associates, Inc.
57 Morehouse Lane
Red Hook, NY 12571 USA
Phone: 845-758-0400
Fax: 845-758-2633
Email: curran@proceedings.com
Web: www.proceedings.com

Volume 2672 Number 43 December 2018

TRR
JOURNAL OF THE TRANSPORTATION RESEARCH BOARD

Contents

Traffic Monitoring: Automobiles, Trucks, Bicycles, and Pedestrians

Articles

TRB

TRANSPORTATION RESEARCH BOARD

CORE SPONSORS

Transportation Departments of the 50 States and the District of Columbia

Federal Government

U.S. Department of Transportation
- Federal Aviation Administration (FAA)
- Federal Highway Administration
- Federal Motor Carrier Safety Administration (FMCSA)
- Federal Railroad Administration (FRA)
- Federal Transit Administration
- National Highway Traffic Safety Administration (NHTSA)
- Office of the Assistant Secretary of Transportation for Research and Technology (OST-R)

U.S. Air Force Civil Engineer Center (AFCEC)
U.S. Coast Guard (USCG)
U.S. Department of Energy
U.S. Department of the Interior (DOI)
U.S. Environmental Protection Agency (EPA)

Nongovernmental Organizations

American Public Transportation Association (APTA)
Association of American Railroads (AAR)
California Air Resources Board

The National Academies of
SCIENCES · ENGINEERING · MEDICINE

The **National Academy of Sciences** was established in 1863 by an Act of Congress, signed by President Lincoln, as a private, nongovernmental institution to advise the nation on issues related to science and technology. Members are elected by their peers for outstanding contributions to research. Dr. Marcia McNutt is president.

The **National Academy of Engineering** was established in 1964 under the charter of the National Academy of Sciences to bring the practices of engineering to advising the nation. Members are elected by their peers for extraordinary contributions to engineering. Dr. C. D. Mote, Jr., is president.

The **National Academy of Medicine** (formerly the Institute of Medicine) was established in 1970 under the charter of the National Academy of Sciences to advise the nation on medical and health issues. Members are elected by their peers for distinguished contributions to medicine and health. Dr. Victor J. Dzau is president.

The three Academies work together as the National Academies of Sciences, Engineering, and Medicine to provide independent, objective analysis and advice to the nation and conduct other activities to solve complex problems and inform public policy decisions. The Academies also encourage education and research, recognize outstanding contributions to knowledge, and increase public understanding in matters of science, engineering, and medicine.

Learn more about the National Academies of Sciences, Engineering, and Medicine at **www.national-academies.org.**

The **Transportation Research Board** is one of seven major programs of the National Academies of Sciences, Engineering, and Medicine. The mission of the Transportation Research Board is to increase the benefits that transportation contributes to society by providing leadership in transportation innovation and progress through research and information exchange, conducted within a setting that is objective, interdisciplinary, and multimodal. The Board's varied activities annually engage about 7,000 engineers, scientists, and other transportation researchers and practitioners from the public and private sectors and academia, all of whom contribute their expertise in the public interest. The program is supported by state transportation departments, federal agencies including the component administrations of the U.S. Department of Transportation, and other organizations and individuals interested in the development of transportation.

Learn more about the Transportation Research Board at **www.TRB.org**.

Article

TRR
JOURNAL OF THE TRANSPORTATION RESEARCH BOARD

Transportation Research Record
2018, Vol. 2672(43) 1–11

DOI: 10.1177/0361198118758391
journals.sagepub.com/home/trr
SAGE

Coupled Approximation of U.S. Driving Speed and Volume Statistics using Spatial Conflation and Temporal Disaggregation

Kartik Kaushik[1], Eric Wood[2], and Jeffrey Gonder[2]

Abstract

The advent of mobile devices with embedded global positioning systems has allowed commercial providers of real-time traffic data to develop highly accurate estimates of network-level vehicle speeds. Traffic speed data have far outpaced the availability and accuracy of real-time traffic volume information. Limited to a relatively small number of permanent and temporary traffic counters in any city, traffic volumes typically only cover a handful of roadways, with inconsistent temporal resolution. This work addressed this data gap by coupling a commercial data set of typical traffic speeds (by roadway and time of week) from TomTom to the U.S. Federal Highway Administration's Highway Performance Monitoring System database of annual average daily traffic (AADT) counts by roadway. This work is technically novel in its solution for establishing a national crosswalk between independent network geometries using spatial conflation and big data techniques. The resulting product is a national data set providing traffic speed and volume estimates under typical conditions for all U.S. roadways with AADT values.

The ubiquitous availability of speed data is revolutionizing the operation of transportation networks, especially roadways. It is now possible to get a bird's-eye view of current conditions on a whole network, spanning a large area (*1*). Rerouting vehicles to avoid congestion and other delays is almost a regular occurrence, especially with advances in route guidance systems built either into vehicles or using smartphones and other mobile devices. Dynamic message signs are increasingly being used to alert travelers of road conditions downstream so that delays are minimized proactively. These advanced traveler information systems are being informed by ubiquitously available speed data (*2*).

Similarly, for planning, a panel of speed data collected over a period helps identify problem areas, bottlenecks, and other issues that may each require different solutions. For example, areas of recurring congestion on certain sections of an urban interstate may prompt a detailed study into the intersection design, the number of lanes, lane widths, demand, and other factors to find workable solutions. Most of these studies, however, require more than just speed data. Such studies rely, perhaps more heavily, on data such as: volume of vehicles during peak hours; mixture of vehicle types; geometric design of the highways; and the rate, type, and severity of incidents.

Researchers and planners are increasingly finding that these studies are made difficult not by the availability of data sets, as most of them are easily acquired, but by the inconsistencies in how different data sets are packaged and reported. No unified underlying geospatial framework exists to which all sets conform, despite global standards and organizations, such as the Open Geospatial Consortium and Technical Committee 211 of the International Organization for Standardization. Private companies in the transportation domain continue to push their own independent geospatial data reporting frameworks or adhere to older, widely used standards (*1*). Public agencies, on the other hand, suffer from a shortage of skilled people capable of bringing existing maps into the modern era (*1*).

For example, the issue tackled here is the spatial and temporal conflation of the Highway Performance Monitoring System (HPMS) volume with Multinet, a commercial base map and traffic speed data set from TomTom Inc., a multinational vendor of maps and other cartographic services. The spatial conflation is complicated by the lack of a unifying framework used by both data sets, which makes join conditions difficult to resolve automatically. The HPMS data set is represented by single line segments that follow the medians of roadways. On the other hand, Multinet is based on a proprietary segmentation scheme that follows the centerlines of

[1]Department of Civil and Environmental Engineering, University of Maryland, College Park, MD
[2]Transportation & Hydrogen Systems Center, National Renewable Energy Laboratory, Golden, CO

Corresponding Author:
Address correspondence to Kartik Kaushik: kartikk@umd.edu

the carriageways. There is no parity in the segment lengths either, and consequently segments from the data sets do not coincide anywhere in the entire United States.

Specific algorithmic tools developed to address these challenges are presented here. The conflation was not perfect, but the average error was under 7%, which is commendable based on other results in the literature for a conflation spanning the entire United States. The developed algorithm can best be described as a scoring algorithm, in which points are rewarded in direct relation to the strength of the match between segment metadata of the two data sets. Most of the metadata were generated using the geometry of the line segments themselves, such as the parallel and perpendicular distances between specific points along the line and the angle of separation between them. Some criteria were borrowed from the few fields that were roughly common between the data sets, such as roadway name and shield numbers.

The remainder of this paper begins with a presentation of available literature on the topic of spatial conflation. This is followed by a detailed description of the base data, including preprocessing steps. Next, spatial join algorithms developed for this paper are presented, including temporal disaggregation of annual average daily traffic (AADT). Finally, results are presented using innovative visualization methods to track the progression of speed with traffic volumes on selected segments.

Literature Review

Conflation, as used here, means a spatial join rather than data fusion, as is commonly used in the literature. Moreover, this conflation can even be viewed as a join in a relational database system, in which joins are used to transfer data from one table or database to another. The spatial join is used to transfer the AADT values from the HPMS data set to the Multinet data set. The literature that deals exclusively with conflation of road geometries covers a wide range of methods, with no one-size-fits-all method. This is perhaps not surprising because each conflation effort has unique challenges and requires a custom-tailored solution.

Road geometry conflation is mainly born out of the need to integrate open access data sets with existing data sets used by private or public agencies. This linkup is increasingly seen as a cost-effective way to keep data sets up to date with information generated by drivers so that they can be used to provide the most applicable real-time solution to issues faced by drivers (*3–5*). The major source of open geospatial data is OpenStreetMaps (OSM), which provides rich details about the real-time urban landscape.

One recent paper published on road geometry conflation uses urban blocks as polygons surrounded by road segments. The goal of the conflation is to connect OSM with government data sets. The polygons from the two sources are matched together in a two-step approach before matching the surrounding road segments. The authors claim that this process has a very low computation cost because the number of urban blocks is fewer than the number of road segments. However, this method is limited to urban areas only (*3*).

Yang et al. proposed a probabilistic relaxation method to again match the OSM data to government data sets. The relaxation method performs a heuristic search of the neighboring space from an initial set created using the dissimilarities between the geometries. The probability matrix is iteratively updated based on selected criteria, until a match is found (*4*). Chen et al. developed an algorithm to attribute floating car data to map segments. The authors argue that existing algorithms found within in-vehicle navigation systems cannot accurately associate low-frequency data to map segments, and propose a dynamic program supported on multiple criteria as a solution (*5*).

Other examples of road geometry conflation in the literature deal with comparing and benchmarking the accuracies of data sets rather than transferring or updating data. Chiefly, these studies also relate to OSM, as it is crucial to understand the caveats in this modern, free, and crowdsourced data set, given the rising importance placed on it. Important studies in this domain are works by Haklay (*6*), Girres and Touya (*7*), and Neis et al. (*8*).

The history of automated road geometry conflation can be traced as far back as 1988 to Saalfeld, who used successive approximations to overlay two map geometries (*9*). Noteworthy papers have since included a statistical approach (*10*) and an approach matching semantics before overlay to ensure the overlay was as accurate as possible (*11*). All these papers used map features, such as lines, points, or nodes, for the conflation. A rule-based method, much like the one proposed here, was first proposed by Cobb et al. (*12*). This method also used the metadata in addition to geometric attributes such as shapes and locations for the conflation.

The decade between 1998 and 2010 is filled with notable contributions by authors using a combination of different data attributes, including geometric and semantic, to achieve a conflation that suits the needs of the study. Altogether, over a 100 relevant papers were produced during this decade (*13, 14*).

Data Description

The goal of the conflation presented here is to create a spatial join between the HPMS and Multinet geospatial data sets to transfer the AADT from the former to the latter. The HPMS public release of geospatial data in shapefiles format is an open, free data set obtained from the Federal Highway Administration Office of Highway Policy Information. The details of this data set are available online (*15*).

TomTom Multinet has two data subsets: a geospatial map layer with road segment geometries and metadata, and a table that provides typical weekly speed profiles on all segments. This is a commercial data set and is available only by purchase. The geospatial data exist in shapefiles and contain the free-flow speed for each segment along with other information, such as road number and name (where available), functional road class (FRC), length, number of lanes, and whether part of a toll road. The typical speed profiles for the week are given as a percentage of the free-flow speeds in the geospatial data sets. These percentages are provided at 5-min epochs in a week, making 288 such epochs in a day.

The typical speed during a 5-min epoch is obtained by the product of the percentage at that epoch and the free-flow speed, for a given day of the week. There are three different free-flow speeds in the geospatial data set for each segment. The first is the speed typical for the whole week, the second is the typical speed during weekdays, and the last is the free-flow speed during weekends. There were minor differences between the three speeds, mainly because these speeds are computed from observed global positioning system probe readings (*1*). It was ensured that the typical profile percentages were multiplied by free-flow speeds from the correct day (i.e., weekday or weekend if available) or the combined speed for the full week when disaggregated weekday and weekend detail is not available. In the version of the data set used in this study, Alaska did not have a profile table; consequently, only a spatial join was performed for Alaska without the disaggregation.

Segmentation Scheme

The HPMS segments were divided either as multiples of the standard length of 160 m (or 0.1 miles) or at points where the AADT values changed significantly. The Multinet segments, on the other hand, were divided using proprietary algorithms such that resolution of the speed data is preserved. The total length of the HPMS network is about 2 billion meters (1.2 million miles), whereas that of the Multinet is 10.6 billion meters (6.6 million miles). The difference is attributed to the vastly greater coverage by Multinet segments, because every road (including the ones within parking lots), and both directions of roadways are mapped.

Figure 1, *a* and *b* show the network from both data sets before the conflation process. Figure 1*a* is a birds-eye view of Denver, Colorado, whereas Figure 1*b* is a zoomed-in section around an interchange. In Figure 1*b*, individual segments are demarked by black dots. The intersection a little southwest from the centers of the maps in Figure 1*b* is I-70 and Denver West Colorado Mills Parkway. I-70 is clearly seen represented using bidirectional segments in the HPMS map, whereas Multinet shows each direction of travel on I-70 using separate one-way segments. Southwest of the intersection is a huge shopping complex; the dense network of Multinet segments show the rows of parking lots. The figures place a perspective on the details captured in the Multinet data set.

Figure 1*b* also explains the major difficulty in the conflation process. North of the intersection is Denver West Parkway, a road parallel to the interstate. The conflation should ensure that the Denver West Parkway segments are joint to the corresponding segments from HPMS and not to the Interstate segments, and vice versa. The developed algorithm of computing parallel and perpendicular distances between the segments scores them based on closeness to address this difficulty. However, some preprocessing was required before applying the algorithm, as explained below.

Common Fields

There are common fields in the two data sets, such as road numbers and names. However, the information and its representation format varied widely. For example, the Multinet data might include all shield numbers from a segment on a road with multiple numbers, whereas the HPMS might just include one, or even none, partly because the segments in the latter are often determined by a change in the AADT values, and roads with other shield numbers may merge or diverge within a segment. The shield numbers may themselves have different notation, such as the inclusion of hyphens, spaces, back or front slash, or other separators to separate the class identifier and the actual road number (e.g., I-95, I95, I 95, I/95, etc.). The most difficult problem, however, was the lack of road names and shield numbers altogether for many of the segments.

The data sets also encoded whether a segment was part of a toll road. This field was useful for an accurate join; however, the road miles that are tolled are insignificantly small as compared to the total length of roads. Another field recorded the number of lanes on the highway, but the count for HPMS included both directions of travel whereas Multinet counted the lanes on one carriageway only. On roads with an equal number of lanes in both directions, this is potentially not a problem; however, on one-way streets or roads with uneven number of lanes, there is no clean way to match the lane counts from both data sets.

The last common field is the FRC. However, both data sets use completely independent classification systems. The HPMS uses FRC values from 1 to 7 whereas Multinet represents FRC from 0 to 11. Often, even the same road may have different FRC values. For example, all grade-separated highways are FRC 1 in the HPMS whereas in Multinet, they may take values from 0 to 2, depending on whether the highway is an interstate, a national U.S. road, or a state highway, and the length for which it is grade separated.

Data Preprocessing

The HPMS preprocessing step split long segments into shorter components to harmonize the geometric distance computations. On the other hand, FRC classes 7 and above (lower in significance than minor collectors) were removed from the Multinet data set in addition to filtering out ferries and other segments that were not part of roads. The following subsections detail the HPMS preprocessing, creation of a geospatial index, and computations of the absolute angles of

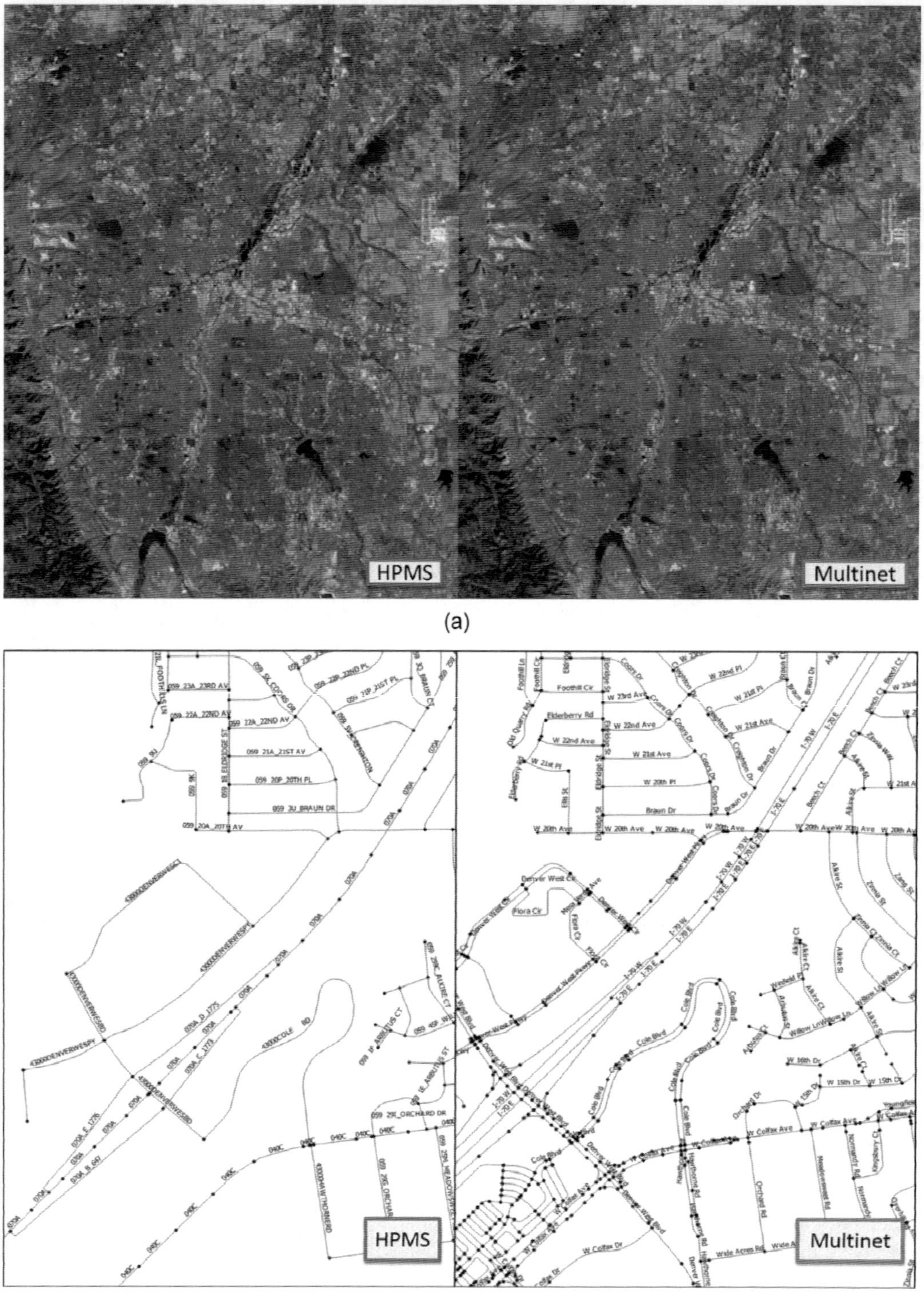

Figure 1. Bird's-eye views of (*a*) Denver, Colorado, as rendered by geometries in the HPMS and Multinet data sets and (*b*) a small area in Denver, Colorado, zoomed in to show geometry details.

the segments with respect to the equator. The angles are used to compute the parallel and perpendicular distances between segments from the two data sets.

Splitting HPMS Segments. Not only did the lengths of the HPMS segments vary widely, but the average segment length was also very different from the mean Multinet segment length. As the conflation algorithm mainly depends on calculating the spatial separation between the segments, having coherent segment lengths is of paramount importance. As the difference in lengths of two adjacent segments increases, the distance between their centroids can take a larger range of values, reducing confidence in eliminating bad join candidates.

For example, consider a hypothetical HPMS segment, which is about a kilometer long (0.6 miles). Adjacent to this segment are tiny (by comparison) Multinet segments, each 100 m long (328 ft). The distance between the centroids of the HPMS segment and the Multinet segment can vary between 0 m (0 ft) near the center and 550 m (1,804 ft) near the ends, assuming minimum overlap. This is a large difference in lengths for a typical road network, as the road can curve away and intersect with other roads within that distance. Therefore, the number of join candidates increases to the point where selection becomes difficult, even impossible.

Consequently, the best strategy was to split the HPMS segments into smaller pieces. All HPMS segments longer than 200 m (656 ft) were selected and broken down into segments 160 m (528 ft) long, which is the standard U.S. chain length, and forms the basis of the HPMS map geometries. HPMS segment lengths, consequently, tend to cluster at multiples of this length. Some states merged these chain-length segments together, only splitting them when the AADT values changed. Splitting at 160 m also reduced the occurrence of very short segments formed by the left-over bits, which reduces computational burden. The splitting cut-off was derived from a histogram of the Multinet segment lengths, which peaked at 200 m.

Longer line segments are often stored as multilines in map geometries, which are just a collection of shorter, regular line segments. The shorter lines encode a start and an end, between which the segments are usually rendered as straight lines. This allows a line to easily follow a curve in a piecewise approximation. To preserve the geometry, the splitting algorithm first exploded all multilines into constituent segments. Any constituent segments longer than 200 m were then subdivided into 160 m segments by linearly interpolating between the start and end coordinates. Vincenty's formula for computing distances on geodesics was used to compute all lengths throughout this paper (*16*). Vincenty's method works better when points are very close together.

The last two digits of the HPMS identification code are the state Federal Information Processing Standards (FIPS) code. The rest of the code can repeat in different states. Therefore, to create a unique identification code after splitting, a three-digit number (with zero-padding as required) was inserted before the FIPS code. The three digits were a linear count of the subsections created from a given HPMS segment. Some segments were split into more than 200 sections. Hereafter, HPMS refers to the new database formed by the split HPMS segments.

Creating a Geospatial Index. A geospatial index serves to limit the search space for matching segments. For example, to join segments on the west side of Kansas, segments on the east side need not be searched. The index was created by applying a buffer to the HPMS segments and storing the buffer extremities in an R-tree (*17*). R-trees work by bounding smaller boxes within larger boxes, similar to real-world addresses. During a lookup, the R-tree is traversed until all the buffers that enclose a provided point are identified. The buffer around HPMS segments was fixed at 500 m (1,640 ft), which in hindsight is a very large distance. A buffer of 100 m (328 ft) would have been sufficient and would have significantly reduced the runtime of the complete algorithm.

The goal of the conflation was to create a one-to-one join between each Multinet segment to the best candidate HPMS segment. Therefore, the algorithm exhaustively combs each valid Multinet segment to find the ideal join partner. However, the join was not forced; without an ideal HPMS segment, the Multinet segment was dropped from the join. As the lookup was from Multinet to HPMS, the geospatial index was created only for the HPMS data set. Candidate HPMS segments were selected by querying the index, and join strength was determined only for selected segments.

Computing Multinet Typical Speeds. The profile speeds were used for the temporal disaggregation of the AADT values transferred through the spatial join. Speeds were attributed to each Multinet segment through a simple relational database join between the geospatial data set and the typical week profile tables. The latter tables repeat each Multinet segment ID seven times, once for each day of the week. Therefore, the join was pivoted so the appropriate free-flow speed column was joined to the correct day: weekday free-flow speeds to weekday profile rows, and for weekends, given a Multinet segment. If the speeds were unavailable for a weekday or weekend day, the generic week free-flow speed was used.

Methodology

A scoring algorithm was used to select the best match for the spatial conflation. As no common fields exist between the data sets, it was not possible to achieve quality conflation just by relying on the metadata for each segment. Further, it was not possible to use the geometries directly because of differences in the way the road segments were encoded. Consequently, conflation was performed by computing additional geometric details about the line segments.

Appendix A of the Mobility Scorecard (*18*) was referenced for disaggregating the AADT by time of day. The steps

Table 1. Scoring Algorithm Tests and Points

Test metric	Test criteria	Points awarded
	Computed geometric data matching criteria	
Perpendicular distance between HPMS and Multinet segments	Distance ≤ 5 m (16 ft)	250
	Distance ≤ 10 m (33 ft)	100
	Distance ≤ 15 m (49 ft)	10
Parallel distance between HPMS and Multinet segments	Distance ≤ 20% of combined HPMS and Multinet segment lengths	100
	Distance ≤ 30% of combined HPMS and Multinet segment lengths	50
	Distance ≤ 50% of combined HPMS and Multinet segment lengths	10
Angle between HPMS and Multinet segments	Angle ≤ 0.01 rad (0.6°)	100
	Angle ≤ 0.1 rad (5.7°)	50
	Angle ≤ 1 rad (57°)	10
	Field metadata matching criteria	
Road shield number of the HPMS and Multinet segments	Road number is the same	1,000
Difference in FRC class between HPMS and Multinet segments	Difference = 0	100
	Difference ≤ 1 class	50
	Difference ≤2 classes	10
Number of lanes in HPMS and Multinet segments	Multinet lanes = half of HPMS lanes	300
Toll road indicator in HPMS and Multinet segments	Flagged as toll roads	300
Threshold to select candidate HPMS segment	Total points should be ≥	320

in this appendix were followed exactly, including adjusting the AADT for the day of the week and computing the peak directions and the severity of congestion. The graphs shown in Exhibits A-2 to A-6 were digitized and used to extract the percentage daily volume for a given hour of the day for scenarios applicable to the respective exhibits. These percentage points were linearly interpolated to obtain the percentage daily volume for 15-min periods in a day, which were then multiplied by the daily volumes obtained by adjusting the directional AADT for the day of the week.

Computing Geometric Data

The Multinet segments were devolved into five points: 0% (start), 25%, 50% (mid), 75%, and 100% (end) of their lengths. As Multinet segments are unidirectional and have well-defined starts and ends, the intermediate points fall in the correct order. On the other hand, as HPMS segments are bidirectional, the starts, ends, and all other intermediate points can be equivalently interchanged. The only stationary point along HPMS segments are the centroids. Therefore, the HPMS segments were only devolved into their centroids because using other points would have necessitated the computations be repeated assuming HPMS segments in reverse order, which does not apply universally to all segments.

The five Multinet points were used to build a set of candidate HPMS segments. The lengths and angles of the imaginary lines connecting the five points to the centroids of the candidate HPMS segments were computed. The lengths of the imaginary lines are then resolved into components parallel and perpendicular to the HPMS segment using the angles. The resolved lengths from all five points are averaged together for each candidate HPMS segment. Additionally, the angle of the Multinet segment with respect to each candidate HPMS segment is computed. The average distances, along with the angles between the segments are used in attributing a score to each candidate HPMS segment.

Computing Angles. Angles between points are easily computed using the inverse tangent (arctan) formula. The Python programming language was used for this research, with the Numpy, Pandas, Geopandas, SQLAlchemy, and Matplotlib packages (*19*). The inverse tangent function is provided by the Numpy package. The two-argument function (atan2) was chosen over the single argument function (arctan) to obtain the signed angle and avoid numeric infinities and division-by-zero errors when the segments are pointing true north, south, east, or west—that is, when the difference between the start and end latitudes or longitudes is zero (*20*).

All angles were computed with reference to the candidate HPMS segments. Use of HPMS segments was motivated by the bidirectional symmetry of the segments; therefore, the absolute angle can be used. Absolute angles are always between 0 and π radians, ensuring that all distances are positive when resolved into parallel and perpendicular components.

Scoring Algorithm

Scores were given based on the computed geometric values and the similarity between the metadata recorded in other fields for each candidate HPMS segment, given a Multinet segment. Table 1 presents the criteria and the corresponding points. The threshold for selection was 320 points, which means at least two criteria must be met at full capacity.

Note that the points and threshold were fine-tuned by looking at sample joins from a wide range of geographies, including the plains of Kansas, the mountains of Colorado, the coastal roads of Hawaii and California, the density of roads in the Northeast, and the sparsity of roads in the Midwest. The candidate segments scoring above the threshold were sorted in descending order of scores, and ties were broken by secondary and tertiary ascending sorts on the perpendicular and parallel distances. The highest scoring, closest segment was thus selected as the join partner.

Another noteworthy point from Table 1 is the relative weights given to each test metric and criterion. Perpendicular closeness is weighted very heavily, as are having the same number of lanes and being part of a toll road. The points are cumulative, that is, if two segments have a perpendicular distance < 5 m, then the total number of points awarded is 360 (250 + 100 + 10), which automatically crosses the threshold. This and the road shield numbers are the only criteria that allow pairing when met in full.

Complete Algorithm

Preprocessing

1. Split the HPMS segments into shorter segments of 160 m (528 ft).
2. Compute the angles with respect to the equator and the centroids of each HPMS segment and store with the HPMS geometries.
3. Construct bounding buffer boxes around the HPMS segments with a 500 m buffer (reduce to 100 m for optimal performance).
4. Build an R-tree index for HPMS bounding boxes.
5. Filter Multinet segments to remove segments with FRC > 6 or where the segment is not a roadway, for example, ferry paths.
6. Compute the typical profile speeds by day of week using the Multinet geospatial data set and profile speed tables. This will be used in step 9.

Conflation

7. For each Multinet segment from the filtered set:
   a. Devolve the segment into five points at 0%, 25%, 50%, 75%, and 100% along the segment length.
   b. For each devolved point:
      (1) Compile a list of candidate HPMS segments by looking up the point in the R-tree index from step 3; go to step 7.*h* if no candidate HPMS segments are found.
      (2) Compute the angle and length of an imaginary line connecting the point and centroid of each candidate HPMS segment identified in step 7.*b*.(1).
      (3) Subtract the angle computed in 7.*b*.(2) from the absolute angle of the HPMS segment computed in step 2.
      (4) Resolve the length computed in 7.*b*.(2) into components parallel and perpendicular to the HPMS segment using the relative angle computed in 7.*b*.(3).
   c. Find the average parallel and perpendicular distances from 7.*b*.(4) for all points from 7.*a*, given a candidate HPMS segment, to obtain the approximate parallel and perpendicular distances between the Multinet and each candidate HPMS segment.
   d. Compute the points scored by each join pair produced by the Multinet segment and candidate HPMS segment using Table 1.
   e. Filter the join pairs for those that score above the threshold in Table 1; if no pairs qualify, go to step 7.*h*.
   f. Sort the filtered pairs by decreasing points, increasing parallel and increasing perpendicular distances.
   g. Designate the HPMS segment in the first pair from step 7.*f* as the join key.
   h. If any step fails to produce a candidate HPMS segment, attribute NULL as the join key to the Multinet segment.

8. Drop Multinet segment with a NULL join key. Copy the AADT values from HPMS to the Multinet data set through the join.

AADT Disaggregation

9. For each Multinet segment with an AADT value obtained through a successful join:
   a. Halve the transferred AADT value for segments part of dual carriageways by referring to the one-way flag in the Multinet metadata.
   b. Factor the AADT from step 9.*a*, for day of week using Exhibit A-1 from the Mobility Scorecard (*18*).

c. Use the typical speed profile for day of week computed in step 6 for steps 9.*d* and 9.*e* below.
d. Determine the peak direction (a.m., p.m., or equal congestion) for each day of the week as detailed in Step 3 of the Mobility Scorecard (*18*).
e. Determine the severity of congestion with respect to the off-peak speed profiles using Equation A-1 in the Mobility Scorecard.
f. Using ordinate values for time of day from Exhibits A-2 to A-6 in the Mobility Scorecard, and the adjusted AADT values as computed in step 9.*b*, and using the peak direction and severity of congestion as determined by steps 9.*d* and 9.*e* above, find the disaggregated traffic counts for each hour in the week, as detailed in the Mobility Scorecard (*18*).
g. Linearly interpolate the typical volumes obtained for each hour in a day from step 9.*f* to find the volumes at 15-min periods in a day; divide these volumes by 4 to ensure the sum of the volumes for each day remains the same as computed in step 9.*b*.

10. End Algorithm.

This algorithm was run state by state, in a program with parallel processing running 14 states simultaneously. Texas, the state with the largest number of segments, took the most time to finish—slightly over 5 days and 13 h, occupying one whole process. The runtime could have been significantly decreased by reducing the buffer around the HPMS segments. The largest penalties to runtime were incurred in computing the angles and lengths of the segments, which required high-precision floating-point geometry calculations.

Results

Figure 2, *a* and *b* show the same geographical areas as Figure 1, *a* and *b*. The line widths have been scaled in proportion to the AADT of each segment. The similarity between the HPMS and Multinet maps scaled by the transferred AADT is clearly visible over a metropolitan region in Figure 2*a*, and spotlights the prowess of the presented conflation algorithm. Figure 2*b*, when compared with Figure 1*b*, shows that the segments in the parking lot southwest of the intersection have disappeared, along with other segments that did not have a join partner. Further, the line widths of Denver West Parkway, north of the intersection, show that it has correctly been conflated with the HPMS segment and was not attributed to I-70. Additionally, the ramps of the Interstate have also been correctly attributed.

The imperfections become apparent, however, when other segments, parallel but unrelated to the HPMS segments, are thickened by AADT, such as the segment in the northeast corner of the map, which was attributed the AADT from I-70. Similarly, the segments to the west of the incorrect segment also show "bleeding" of AADT to unrelated segments. These are primarily a product of having a single threshold. The threshold was fixed by considering winding roads, and consequently is far too lenient for straight roads. In addition, some Multinet segments are missing around the intersection, indicating the join failed for these segments.

Table 2 presents a global perspective of the errors. The table shows the percentage error in the total length of conflated Multinet segments when compared with the total length of HPMS segments for each state. The final value for the whole country shows that the average error is just under 7%, which for 1.2 million miles of complex, real-world roadways is very good by standards in the literature.

Figure 3 shows the use of the conflated data set. This figure shows the relation of the speed and volume data sets for I-405N in Los Angeles, California, and I-270N in Maryland. Figure 3, *a* and *c* are created by plotting the typical week speed profiles by day, and thickening the lines by the traffic volume for each 15-min period of that day. Figure 3, *b* and *d* use arrows to present the evolution of traffic congestion with respect to time, volume, and speeds. These figures also share a marked resemblance with speed–flow fundamental diagrams.

Conclusion

A scoring algorithm for conflation was proposed here and was applied to the entire U.S. network. The total conflation error was found to be less than 7%, which is commendable when compared with other results in the literature. For example, the probabilistic relaxation method proposed by Yang et al. has an error of about 5% for a much smaller urban study area (*4*). By constraining the area, the types of geographies are limited, and therefore the parameters of Table 1 can be further tuned for best results. However, the United States has wide ranging and vastly different geographies, which makes using a generic conflation algorithm difficult.

The data set created by the conflation is being used by the National Renewable Energy Laboratory to answer specific questions about the effects of factors such as congestion, volume of traffic, gradient of roadways, and weather conditions on fuel use and greenhouse gas emissions. The plots in Figure 3 present an innovative way to visualize the conflated data and were produced to aid one such study conducted at the National Renewable Energy Laboratory.

A possible avenue for improvement of the conflation algorithm is to make the scores and thresholds more sensitive and directly dependent on the attributes of the segments. Additionally, successful join candidates can also be used to train advanced deep neural networks for better joins and to help resolve the issues with the current conflated data set.

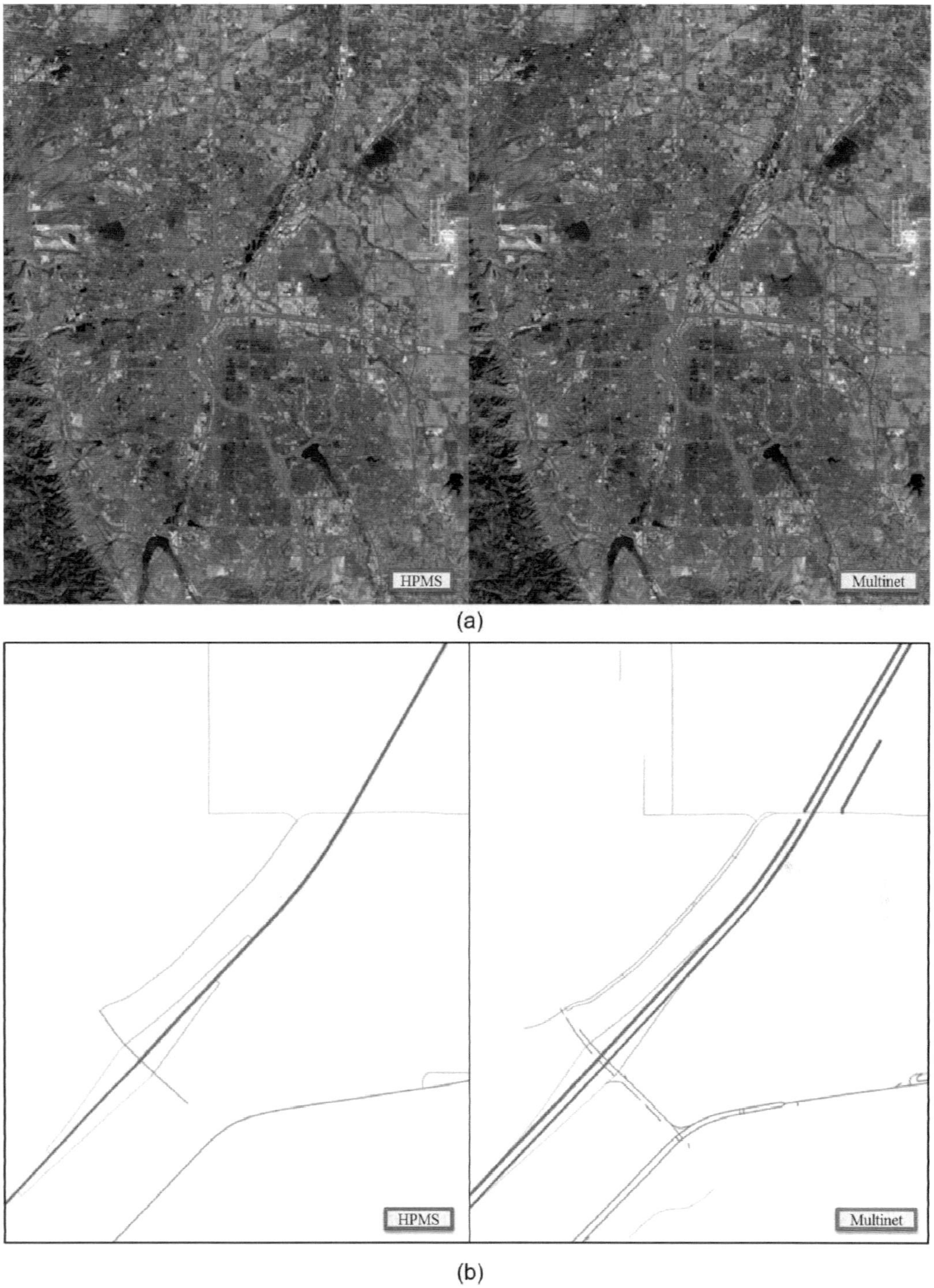

Figure 2. Bird's-eye views of (*a*) Denver, Colorado, post conflation, with lines thickened in proportion to AADT values and (*b*) an interchange in Denver, Colorado (zoomed-in), with lines thickened to show AADT values.

Table 2. Total Absolute Conflation Error by State and for the Whole Country

State FIPS	Total error	State FIPS	Total error	State FIPS	Total error
01	0.044	21	0.080	38	0.033
02	−0.254	22	0.307	39	0.147
04	0.207	23	0.014	40	0.171
05	0.055	24	0.178	41	−0.027
06	0.282	25	0.103	42	0.057
08	0.141	26	0.123	44	0.183
09	0.046	27	0.124	45	−0.310
10	0.004	28	0.075	46	0.083
11	−0.109	29	0.153	47	0.108
12	0.475	30	−0.475	48	0.278
13	0.089	31	0.020	49	−0.068
15	0.156	32	0.051	50	−0.064
16	−0.008	33	0.024	51	−0.291
17	0.230	34	0.347	53	−0.031
18	0.139	35	0.263	54	−0.439
19	0.163	36	0.200	55	0.130
20	0.056	37	0.087	56	0.129
Country average absolute error				0.068	

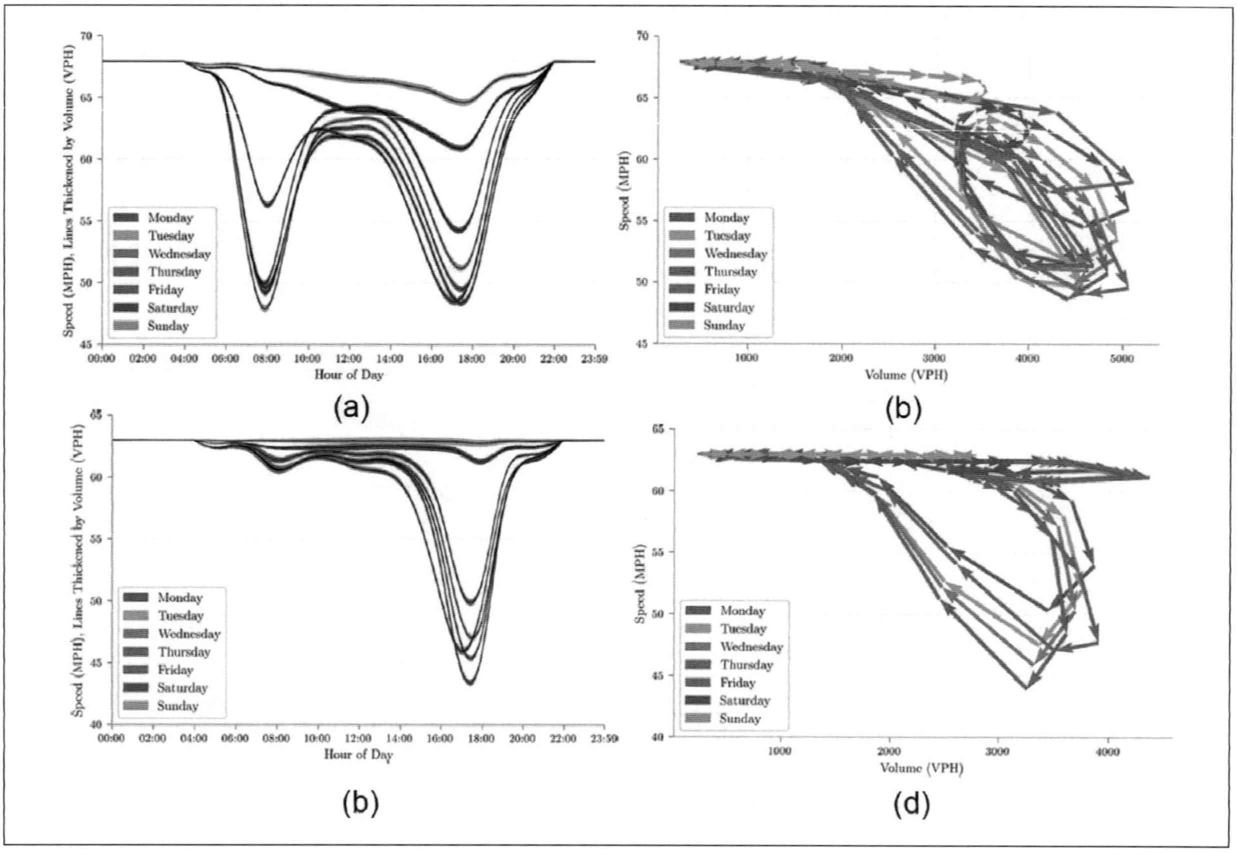

Figure 3. (*a*) Speed and volume on I-405N in California for a typical week; (*b*) evolution of congestion during a typical week on I-405N in California; (*c*) speed and volume on I-270N in Maryland for a typical week; (*d*) evolution of congestion during a typical week on I-270N in Maryland.

Acknowledgments

This work was supported by the U.S. Department of Energy under Contract No. DE-AC36-08GO28308 with Alliance for Sustainable Energy, LLC, the Manager and Operator of the National Renewable Energy Laboratory. Funding provided by the U.S. Department of Energy Office of Energy Efficiency and Renewable Energy Vehicle Technologies Office. The U.S. Government retains and the publisher, by accepting the article for publication, acknowledges that the U.S. Government retains a nonexclusive, paid-up, irrevocable, worldwide license to publish or reproduce the published form of this work, or allow others to do so, for U.S. Government purposes.

References

1. Kaushik, K., E. Sharifi, and S. E. Young. Computing Performance Measures with National Performance Management Research Data Set. *Transportation Research Record: Journal of the Transportation Research Board*, 2015. 2529: 10–26.
2. Zhang, Y. *Uncertainty Associated with Travel Time Prediction: Advanced Volatility Approaches and Ensemble Methods*. PhD dissertation. Department of Civil and Environmental Engineering, University of Maryland, College Park, 2015.
3. Fan, H., B. Yang, A. Zipf, and A. Rousell. A Polygon-Based Approach for Matching OpenStreetMap Road Network with Regional Transit Authority Data. *International Journal of Geographical Information Science*, Vol. 30, No. 4, 2016, pp. 748–764. http://dx.doi.org/10.1080/13658816.2015.1100732.
4. Yang, B., Y. Zhang, and X. Luan. A Probabilistic Relaxation Approach for Matching Road Networks. *International Journal of Geographical Information Science*, Vol. 27, No. 2, 2013, pp. 319–338. http://dx.doi.org/10.1080/13658816.2012.683486.
5. Chen, B. Y., H. Yuan, Q. Li, W. H. K. Lam, S. L. Shaw, and K. Yan. Map-Matching Algorithm for Large-Scale Low-Frequency Floating Car Data. *International Journal of Geographical Information Science*, Vol. 28, No. 1, 2014, pp. 22–38. http://dx.doi.org/10.1080/13658816.2013.816427.
6. Haklay, M. How Good is Volunteered Geographical Information? A Comparative Study of OpenStreetMap and Ordnance Survey Datasets. *Environment and Planning B: Planning and Design*, Vol. 37, No. 4, 2010, pp. 682–703. http://dx.doi.org/10.1068/b35097.
7. Girres, J. F., and G. Touya. Quality Assessment of the French OpenStreetMap Dataset. *Transactions in GIS*, Vol. 14, No. 4, 2010, pp. 435–459. http://dx.doi.org/10.1111/j.1467-9671.2010.01203.x.
8. Neis, P., D. Zielstra, and A. Zipf. The Street Network Evolution of Crowdsourced Maps: OpenStreetMap in Germany 2007–2011. *Future Internet*, Vol. 4, No. 1, 2011, pp. 1–21.
9. Saalfeld, A. Conflation: Automated Map Compilation. *International Journal of Geographical Information System*, Vol. 2, No. 3, 1988, pp. 217–228. http://dx.doi.org/10.1080/02693798808927897.
10. Walter, V., and D. Fritsch. Matching Spatial Data Sets: A Statistical Approach. *International Journal of Geographical Information Science*, Vol. 13, No. 5, 1999, pp. 445–473. http://dx.doi.org/10.1080/136588199241157.
11. Ware, J. M., and C. B. Jones. Matching and Aligning Features in Overlayed Coverages. *Proc., 6th ACM International Symposium on Advances in Geographic Information Systems*, New York, NY, 1998, pp. 28–33.
12. Cobb, M. A., M. J. Chung, H. Foley, III, F. E. Petry, K. B. Shaw, and H. V. Miller. A Rule-Based Approach for the Conflation of Attributed Vector Data. *GeoInformatica*, Vol. 2, No. 1, 1998, pp. 7–35.
13. Ruiz, J. J., F. J. Ariza, M. A. Ureña, and E. B. Blázquez. Digital Map Conflation: A Review of the Process and a Proposal for Classification. *International Journal of Geographical Information Science*, Vol. 25, No. 9, 2011, pp. 1439–1466. http://dx.doi.org/10.1080/13658816.2010.519707.
14. Yuan, S., and C. Tao. Development of Conflation Components. *Proc., Geoinformatics*, Ann Arbor, 1999, pp. 1–13.
15. Office of Highway Information Management. *Highway Performance Monitoring System (HPMS)*. Federal Highway Administration (FHWA) FHWA-PL-98-045, 1998.
16. Vincenty, T. Direct and Inverse Solutions of Geodesics on the Ellipsoid with Application of Nested Equations. *Survey Review*, Vol. 23, No. 176, 1975, pp. 88–93. http://dx.doi.org/10.1179/sre.1975.23.176.88.
17. Guttman, A. R-Trees: A Dynamic Index Structure for Spatial Searching. *ACM*, Vol. 14, No. 2, 1984, pp. 47–57.
18. Schrank, D., B. Eisele, T. Lomax, and J. Bak. *Appendix A: Methodology for the 2015 Urban Mobility Scorecard*. Texas A&M University, Texas, 2015.
19. Millman, K. J., and M. Aivazis. Python for Scientists and Engineers. *Computing in Science & Engineering*, Vol. 13, No. 2, 2011, pp. 9–12.
20. Organick, E. I. *A Fortran IV Primer*. Addison-Wesley Publishing Co., Inc., Boston, MA, 1966.

The Standing Committee on Highway Traffic Monitoring (ABJ35) peer-reviewed this paper (18-06756).

Article

TRR
JOURNAL OF THE TRANSPORTATION RESEARCH BOARD

Transportation Research Record
2018, Vol. 2672(43) 12–23

DOI: 10.1177/0361198118775840
journals.sagepub.com/home/trr
SAGE

A Vibration-Based Vehicle Classification System using Distributed Optical Sensing Technology

Hongduo Zhao[1], Difei Wu[1], Mengyuan Zeng[1], and Sheng Zhong[1]

Abstract

This paper presents a vibration-based vehicle classification system using distributed optical vibration sensing (DOVS) technology and describes a comprehensive classification method including signal processing and feature extraction. With low maintenance costs, this system can collect vehicle classification data in a larger scale. At first, it utilizes an embedded sensing fiber as a distributed sensor to collect traffic-induced vibration signals, and then extracts several features from the raw signals to estimate axle configurations and identify vehicle categories. At the same time, an empirical mode decomposition (EMD)-based method is applied to reconstruct signals for features extraction, and then several extraction algorithms are proposed to obtain the axle configuration, moving speed, and frequency-domain feature of each vehicle. When all features are extracted, a multi-step classifier is designed to categorize vehicles into different classes. In addition, to evaluate the classification performance of this system, a prototype system was installed on a relief road in Shanghai, China using precast concrete pavement technology. With an overall accuracy of 89%, the test results show a good performance of this classification system.

Vehicle classification data plays a significant role in traffic and road management. Traffic management can use vehicle type information for road toll, traffic control, and traffic guidance. Additionally, the information of trucks can be used to assess the performance of past and current pavement as well as to predict future performance, thus offering important assistance to road management.

Various technologies have been developed for vehicle classification over recent decades and these are broadly divided into two categories: non-embedded and embedded detectors. The non-embedded detectors, including vision-based detectors (*1–3*), acoustic-based detectors (*4*), and laser-based detectors, are installed beside or over the road. With this type of detector, pavement cuts or lane closures are not required during installation and precise information about vehicle types can be obtained. However, the performance of the classification is sensitive to environmental conditions, such as visibility, noise level, and temperature. By contrast, embedded detectors such as inductive loops (*5*), magnetometers (*6*), and piezoelectric sensors (*7*) are installed beneath the pavement surface. They are not sensitive to environmental conditions but there is often a high cost in the installation and maintenance.

Recently, vibration-based methods for traffic monitoring have become a more efficient and less costly. One of their advantages is that they are able not only to obtain traffic parameters from in-pavement detectors but also to monitor pavement damage and loading conditions, which is beneficial to both traffic and road management. In the field of vibration monitoring, the accelerometer is the most commonly used detector. Some accelerometer-based systems and algorithms have been proposed for axle detection and vehicle classification. Bajwa et al developed an accelerometer-based wireless sensor network for vehicle classification and also proposed a novel algorithm to count axles and estimate axle spacing (*8*). Using accelerometers and magnetometers, Ma et al developed a wireless automatic vehicle classification prototype system and used a filter algorithm to identify axles (*4*). Additionally, Yeum et al used a Viola-Jones algorithm to extract and classify the distinctive dynamic patterns of different vehicles by converting the measured acceleration signals to time-frequency images (*9*). Markus and Hostettler also used the frequency-domain features to detect and classify vehicles (*10*).

[1]Key Laboratory of Road and Traffic Engineering of the Ministry of Education, Tongji University, Shanghai, P. R. China

Corresponding Author:
Address correspondence to Difei Wu: wdf_michael@163.com

Although the accelerometer-based systems can offer high-precision information, they are not suitable for large-scale measurement, and may require high-cost maintenance. The distributed optical vibration sensing-system (DOVS), which is an emerging technology in the sensing field, has attracted increasing attentions in laboratory studies and engineering applications. DOVS has numerous advantages such as large-scale measuring, resistance to damage, anti-electromagnetic interference, and highly reliable performance in severe environments (*11*). Moreover, DOVS is a distributed sensing system because every single unit in the fiber can be used as a sensor. Therefore, DOVS is a low-cost and easy-to-implement system compared with single-unit sensors, and there is a great potential for detecting and classifying large-scale vehicles based on DOVS.

This paper presents a vibration-based vehicle classification system with the application of DOVS technology. This system is capable of counting the number of axles, estimating the axle spacing, extracting frequency-domain features, and classifying vehicles. The remainder of this paper is organized as follows: the operating principle of DOVS and the prototype system is introduced in Section 2 along with the installation method of the classification system; the signal processing methods are proposed in Section 3; a series of feature extraction approaches are proposed and described in Section 4; after the extraction of axle features and frequency-domain feature, a multi-step classifier is designed and presented in Section 5; the conclusions are presented in the last section.

System Description

Operating Principle of DOVS

DOVS system is a distributed sensing system that can obtain the vibration information along the sensing fiber. The basic principle of vibration measurement relies on the phase optical time-domain reflectometer (ϕ-OTDR) technology, which was first proposed in 1993 (*12*). Figure 1 shows a typical ϕ-OTDR system architecture. The system uses a laser source with narrow linewidth and low frequency shift to form the interference of the back-scattered light waves. The measurement trace of the system shows the intensities of the Rayleigh Backscattered light generated at different positions in the sensing fiber.

The output intensity (I) of the backscattered light is related to the optical phase in the sensing fiber (*13*):

$$I = I_1(t) + I_2(t) + 2\sqrt{I_1(t)I_2(t)}\cos(2\varphi + \varphi_0) \quad (1)$$

where

$I_1(t)$ = the intensity of the first interference light,

$I_2(t)$ = the intensity of the second interference light,

φ = the optical phase of backscattered light at a certain position,

φ_0 = the phase difference between the two interference lights.

When external vibration is applied at certain position, the corresponding phase of backscattered light will change, resulting in an oscillation of the output intensity. Therefore, the DOVS system can convert the external

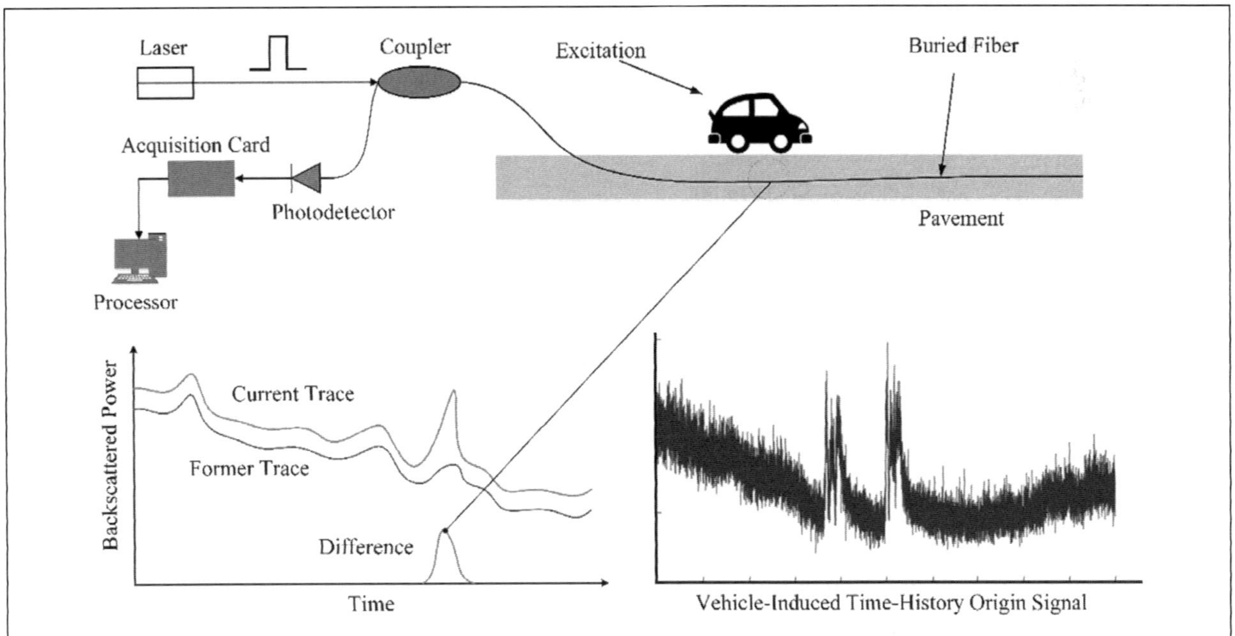

Figure 1. Architecture of DOVS.

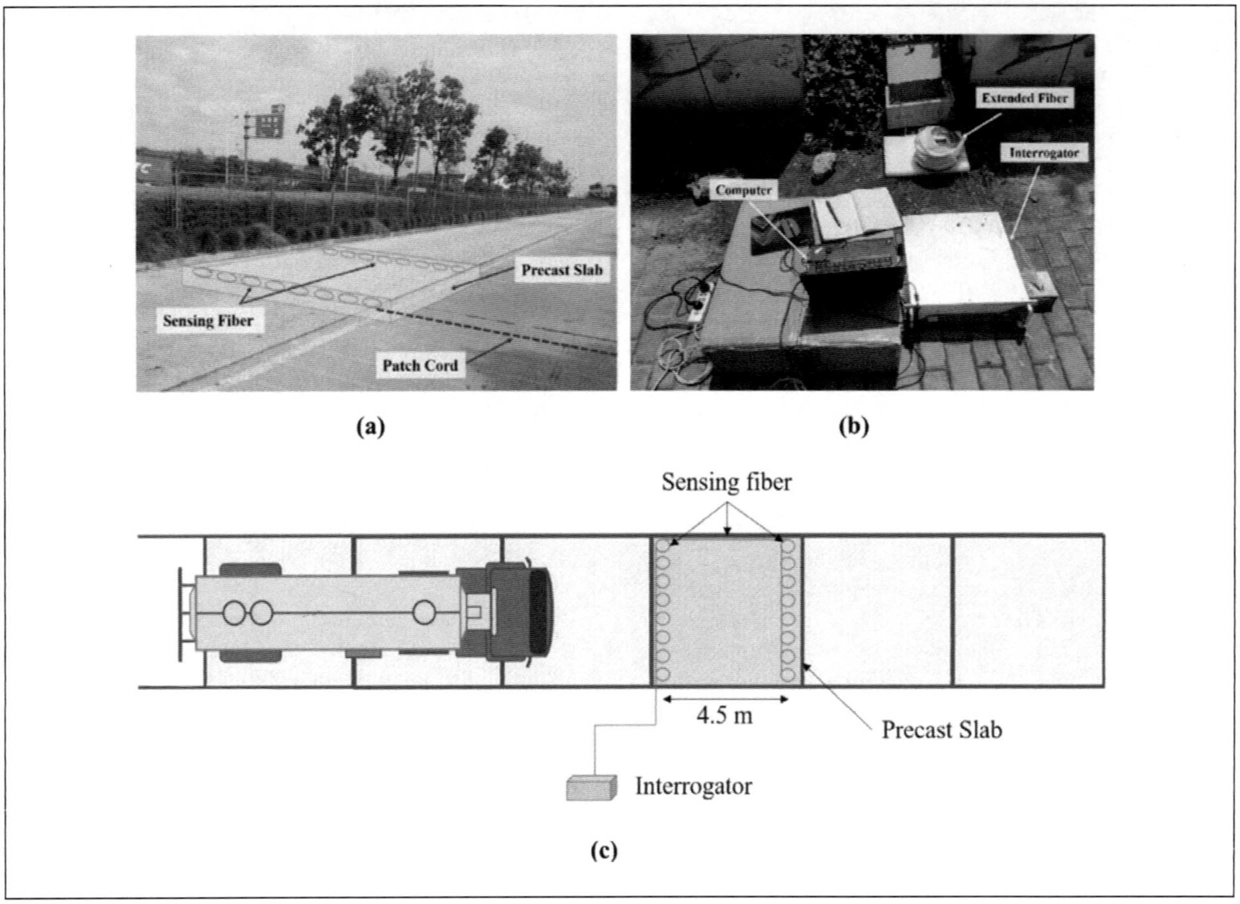

Figure 2. (*a*) Test site; (*b*) measurement devices; (*c*) system set-up.

vibration to this oscillation using a coupler and a photodetector, and then record them by an acquisition card and a processor. Moreover, the position of the external vibration is represented by the distance from the start, which is determined by calculating the time delay between the injected pulsed light and backscattered light. The positioning accuracy of the system (Δz), namely spatial resolution, relates to the pulse width (T_p) of the pulsed light:

$$\Delta z = \frac{cT_p}{2n} \tag{2}$$

where
c = the velocity of light in vacuum,
n = the refractive index of the sensing fiber.

System Components

Figure 2 illustrates the test site, measurement devices and a scheme of the prototype system. There are three major components in the system including sensing fiber (3 mm-diameter, single mode), interrogator (BAIANTEK, FT630) and a computer.

The sensing fiber is buried beneath the concrete pavement surface, and coated with a steel wire and a polyethylene sheath, which can protect the fiber from being damaged during installation and operation. The vibration and position information are acquired through the interrogator and the computer, with a sampling frequency of 2000 Hz and a spatial sampling interval of 2 m. Moreover, the embedded sensing fiber is firstly circled to sixteen fiber coils (30 cm diameter, four rings) with adequate spaces, and then arranged into two rows with a 4.5 m space (Figure 2*a* and *c*). Through increasing the layout density, this layout improves the accuracy of spatial localization.

System Installation

To test and verify the system, a prototype system was installed on a relief road in Shanghai to collect the traffic-induced vibration signals. The traffic composition was complex, including cars, buses, trucks, and multi-axle trailers.

The installation of conventional in-pavement sensors such as induction loops and magnetic sensors is

Table 1. Vehicle Classification Categories

Class	Vehicle type	Axle composition	Number
1	Passenger car/van	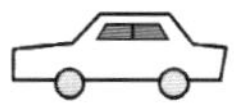	247
2	Two-axle bus	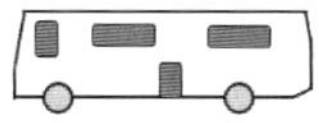	32
3	Two-axle truck	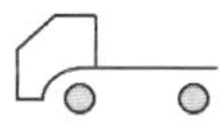	119
4	Three-axle truck		10
5	Three-axle truck		1
6	Four-axle truck		10
7	Four-axle trailer	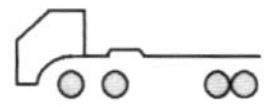	50
8	Five-axle trailer	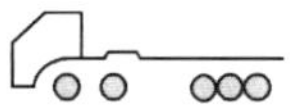	19
9	Five-axle trailer	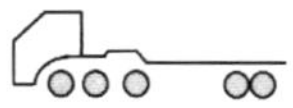	1
10	Six-axle trailer	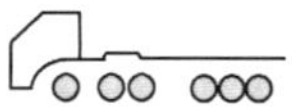	3
Total			482

destructive and may decrease pavement life. For the prototype system, the precast concrete pavement technology is utilized to avoid these problems. Firstly, the sensing fiber was buried into a precast concrete slab when casting and the end of the fiber was placed at the corner of the slab. When the slab was assembled, the end was taken out and connected to the interrogator, which completed the system installation. The installation procedures were carried out in four hours.

Additionally, to improve the positioning performance of the system, a 10 km-length extended fiber is utilized to connect the sensing fiber to the interrogator (Figure 2*b*).

Data Acquisition

Vibration signals induced by over 400 vehicles were collected with this system. The speeds of the vehicles ranged between 8 and 25 m/s. The vehicles were classified into 10 categories according to their axle features and usage (passenger or freight transport), the descriptions and number of each category are presented in Table 1.

Signal Processing

Empirical Mode Decomposition

The raw vibration signal was complicated as it consisted of noisy signal and useful signal. For axle-induced vibration signals, the useful signal mainly concentrates on specific frequency bands while noisy signals are evenly distributed over frequency. To eliminate noises and extract useful signals, the empirical mode decomposition (EMD) method is utilized for signal reconstruction. It is a decomposition method that can decompose raw signal into several intrinsic components, namely, intrinsic mode functions (IMFs). Each IMF represents the detail components of the raw signal of different frequency bands.

An IMF is a function that satisfies the following two requirements: (1) the number of extrema and the number of zero crossings must either be equal or differ at most by one; (2) the running mean value of the envelope defined by the local maxima and the envelope defined by the local minima is zero. The procedures to decompose the raw data ($f(t)$) are shown as follows (*14*):

Firstly, the local extrema $x(t)$ are identified. The local maxima are connected forming the upper envelope $u(t)$ while the local minima are connected forming the lower envelope $l(t)$. This connection is implemented by a cubic spline interpolation scheme. Then, the running mean can be defined as:

$$m_1(t) = \frac{u(t) + l(t)}{2} \tag{3}$$

Then, the first component $h_1(t)$ is computed by subtracting $m_1(t)$ from the signal $x(t)$:

$$h_1(t) = x(t) - m_1(t) \tag{4}$$

The component will be checked if it meets the two IMF requirements. If not, $h_1(t)$ is treated as the new data, and the steps above will be repeated until the component becomes an IMF. Then, the first IMF is obtained, which is denoted with $C_1(t)$.

Next, we define the residue:

$$r_1(t) = x(t) - C_1(t) \tag{5}$$

The residue $r_1(t)$ is considered as the new signal and the sifting process is applied from the beginning. As a result, the signal $x(t)$ will be decomposed into a finite number of IMFs $C_i(t)$. The sifting process ends when the last residue $r_n(t)$ is a constant or a monotonic function. By summing up all the IMFs and the final residue, the $x(t)$ is written as:

$$x(t) = \sum_{i=1}^{n} C_i(t) + r_n(t) \tag{6}$$

Signal Denoising and Reconstruction

Figure 3 illustrates the decomposition of a vibration signal induced by a four-axle truck and two passenger cars. The first time-amplitude curve shows the raw signal. Obviously, IMFs in different frequency bands show different patterns: in high-frequency IMFs (IMF1~2), the vehicle-induced signals are not significant; in IMF 3~5, the axle configuration is easy to identify. In the low-frequency IMFs (IMF 7~9), the vehicle-induced vibration is significant. However, the axle configuration can hardly be identified due to the signal superposition. This indicates that the superposed part mainly concentrates on the low frequency band while less of them are superposed in high frequency bands.

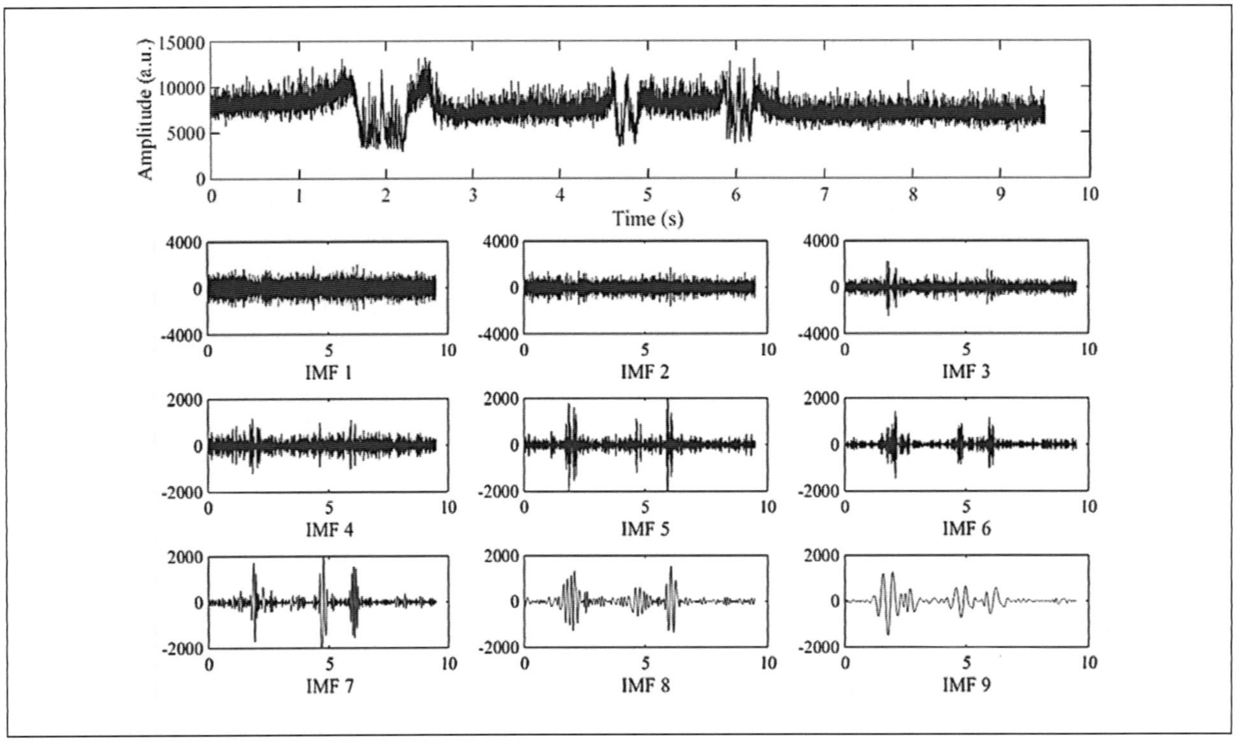

Figure 3. Empirical mode decomposition.

Although the selected IMFs contain useful signal, the noisy signal cannot be neglected. We use a soft threshold function to eliminate the noisy part of selected IMFs:

$$\eta_t(x) = \begin{cases} 0 & |x(i)| > th \\ \mathrm{sgn}(x(i))(|x(i)| - th) & \text{otherwise} \end{cases} \quad (7)$$

where
$\eta_t(x)$ = the denoised signal.
The threshold th is calculated by

$$th = \sigma\sqrt{2\log N} \quad (8)$$

Then, the reconstructed signals are reconstructed by superposing several denoised IMFs (IMF $n\sim m$):

$$fr(t) = \sum_{i=n}^{m} \eta(IMF_i) \quad (9)$$

In this paper, three signals ($fr_1(t)$, $fr_2(t)$, $fr_3(t)$) are reconstructed for three purposes: event extraction, axle feature extraction, and frequency-spectrum analysis. For event extraction, IMF 8~9 are selected to reconstruct the signal; for axle feature extraction, IMF 3~5 are selected; for frequency-spectrum analysis, IMF 1~9 are selected.

Short-Time Energy Analysis

The vehicle-induced vibration signal is time-varying and nonstationary in a long segment. However, signal in a short segment is nearly stationary. Therefore, the signal can be split into many short segments to be processed. In this paper, a short-time energy method is used to split the reconstructed signal into a series of short segments.

The short-time energy of time series $x(m)$ is defined by

$$E_n = \sum_{m=n-N+1}^{n} [x(m)\cdot w(n-m)]^2 \quad (10)$$

where
$E(n)$ = the short-time energy of the nth frame,
$w(n)$ = the window function.
Moreover, the rectangular window function is selected:

$$w(n) = \begin{cases} 1, 0 \le n \le M-1 \\ 0, \text{otherwise} \end{cases} \quad (11)$$

where
M = the width of the rectangular window.
For the vehicles at different speeds, the width can be adjusted for feature extraction.

Features Extraction

Event Extraction

The first step in feature extraction is to identify the moving vehicles and extract the corresponding signals from a long-time series and the reconstructed signal $fr_1(t)$ is utilized in this step.

In the prototype system, this step was conducted every 10 or 20 seconds. Short-time energy analysis, whose window width is 1,000 (equals to 0.5 s), was utilized to determine whether the vehicle was present or not. Each event was identified by deciding if the energies of continuous time segments were significantly high compared with the energies in "no vehicle" segments. Then, the corresponding time series were extracted.

Axle Features Extraction

Axle features include axle number and axle spaces. According to Figure 3, in IMF 4~6, axles will induce several separated short-time vibrations in a long-time series. Therefore, we can use short-time energy analysis to separate each axle from others, and then extract axle features from the time–energy curve. Figure 4 presents

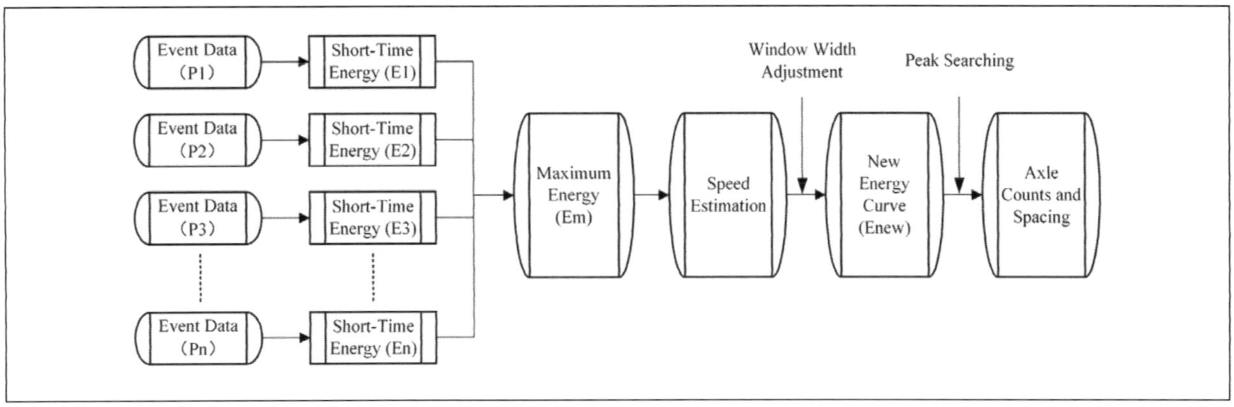

Figure 4. Procedures of axle identification.

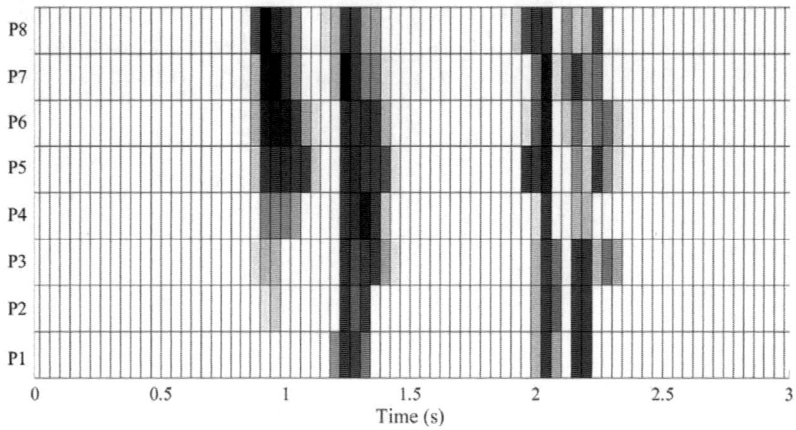

Figure 5. Time–energy distribution of different points.

the procedures of axle feature extraction. In this procedure, the reconstructed signal $fr_2(t)$ is utilized.

Multi-Point Combination

During the test, it was found that the positions of passing wheels have a certain randomness in cross section. If the sampling points are located in the middle of the two-wheel trajectories, the collected signal will be a superposed result of the signals induced by left and right wheels, resulting in difficulties in the axle counting. As a result, a multi-point combination method is proposed to improve the identification accuracy.

As the wheels in a single axle are symmetric, half of the sampling points were selected (eight points) to make the analysis. Figure 5 shows the short-time energies of the selected eight points (a four-axle trailer, the window width is 80); P8 locates on the edge of the slab while P1 locates in the middle of the slab; the color of each grid represents the corresponding short-time energy. Obviously, the four dark bands represent the four axles of the trailer. However, the short-time energies of several points are inadequate for axle identification. For example, the energy of the front axle is lower at P1, P2 and P3, and the energy of the tandem-axle is lower at P4 and P6. To ensure the reliability of identification, the sampling points that are closest to the wheel path should be selected for axle identification. Thus, the maximum instant energy for each sampling point is selected to rebuild a new time series $E_m(n)$:

$$E_m(n) = \max(E_1(n), E_2(n) \cdots E_8(n)) \tag{12}$$

where

$E_1(n), E_2(n) \cdots E_8(n) =$ time–energy curves of P1, P2……P8.

Speed Estimation

The speed of the passing vehicle is an important factor in association with the duration of the traffic-induced vibration. It can be estimated by calculating the ratio of the distance between adjacent measuring positions (ΔL) to the time difference (T) of the vehicle passing over the two positions:

$$v = \frac{\Delta L}{T} \tag{13}$$

The time difference was derived by calculating the difference between the times at which the front axle (most of the front axles are single-axle) moved across the two measuring positions. The two times for exacting were obtained from the maximum time series (E_m) through searching the first peaks.

Axle Identification

Figure 6 shows the time–energy curves (induced by a five-axle truck and a two-axle small truck) with different window widths, suggesting that the window width will affect the axle identification. With a width of 150, the two axles of the two-axle small truck are easily separated from the time–energy curve. However, the two axles in the tandem axle of the five-axle truck cannot be separated while they are identified with a window width of 100. Therefore, the window width needs to be adequately small to reliably identify the rear axles of trucks (tandem axles or tridem axles, whose spacing is usually 1~1.5 m). In this paper, the window width M is determined by

$$M = \frac{\Delta D}{2v} \cdot f_s \tag{14}$$

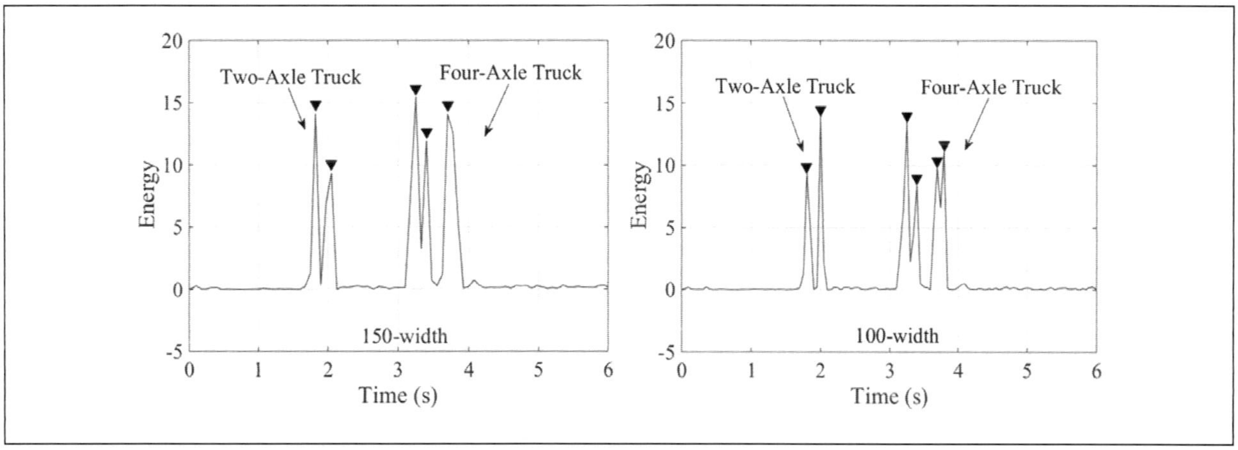

Figure 6. Time–energy curves with different window widths.

where
ΔD = the spacing of the tandem axles or tridem axles,
v = the vehicle speed,
fs = the sample frequency.

Then, the axles are counted by searching the peaks in the Time–energy curve, and the axle spacing can be determined by calculating the product of the time between adjacent peaks and moving speed,

$$d = v \cdot t \tag{15}$$

Generally, vehicle axles are classified into three categories: single axle, tandem axle (two rear axles), and tridem axle (three rear axles). In China, axle spacing of the tandem axle and tridem axle ranges between 1.0 and 1.35 m, while the distance between two adjacent single-axle is greater than 1.8 m. Moreover, for trailers, the space between first two axles ranges between 3.2 and 3.6 m, while the space is usually below 2.0 m in four-axle trucks (the spacing information was obtained from *http://www.chinacar.com.cn/*). Consequently, axle configurations of moving vehicles are easily determined depending on the space calculated between adjacent axles. Subsequently, the multi-axle trucks and trailers are identified based on axle configurations. As shown in Figure 6, it was calculated that the axle spacing of the four-axle truck was 1.9 + 3.9 + 1.3 m, so it was determined that the vehicle included two single-axles and a tandem axle.

Frequency-Domain Feature Extraction

Moving vehicles will not only affect the vibration features in time domain, but also cause a variation of frequency-domain features, such as frequency spectrum. The frequency-domain features of the pavement are influenced by the speeds, weights, and suspension systems. As vehicles of different types have different suspension systems and weights, the frequency analysis can also be used to identify the type of moving vehicles.

To transform the time-domain signal into frequency domain, a short-time Fourier transform (STFT) method is used to describe how the amplitude of a signal is distributed over frequency. Assuming the time-domain signal is $S(\tau)$, the equation of STFT is

$$S_t(\omega) = \sum_{\tau} S(\tau)h(\tau - t)e^{-j\tau\omega} \tag{16}$$

where
$S_t(\omega)$ = the spectrum function,
$h(\tau)$ = the window function,
ω = the angular frequency.

The frequency spectrums of different vehicles (passenger car, bus, truck, and trailer) are presented in Figure 7. The speeds of the four vehicles are 15, 18, 16, 14 m/s, respectively. It shows that there are significant differences between the four vehicles in the frequency domain. For passenger car and truck, the dominant frequencies were mostly concentrated on the low-frequency band (below 30 Hz), while the distributions of bus-induced and trailer-induced signals were wider in frequency domain.

As the induced signals of four types of vehicle are distributed differently in the frequency domain, we extracted the "weighted frequency" as an index to characterize the amplitude distribution in the frequency domain, which is calculated by:

$$f_w = \frac{\int_{f_l}^{f_h} S(f) \cdot f df}{\int_{f_l}^{f_h} S(f) df} \tag{17}$$

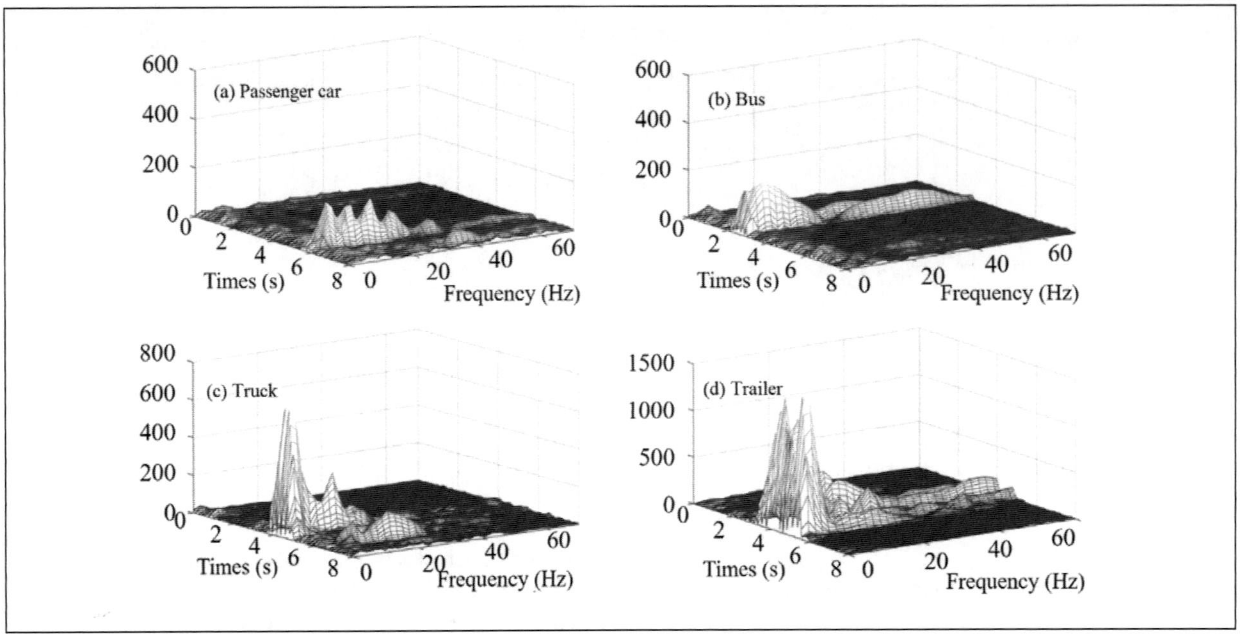

Figure 7. Frequency spectrums of different vehicles over time.

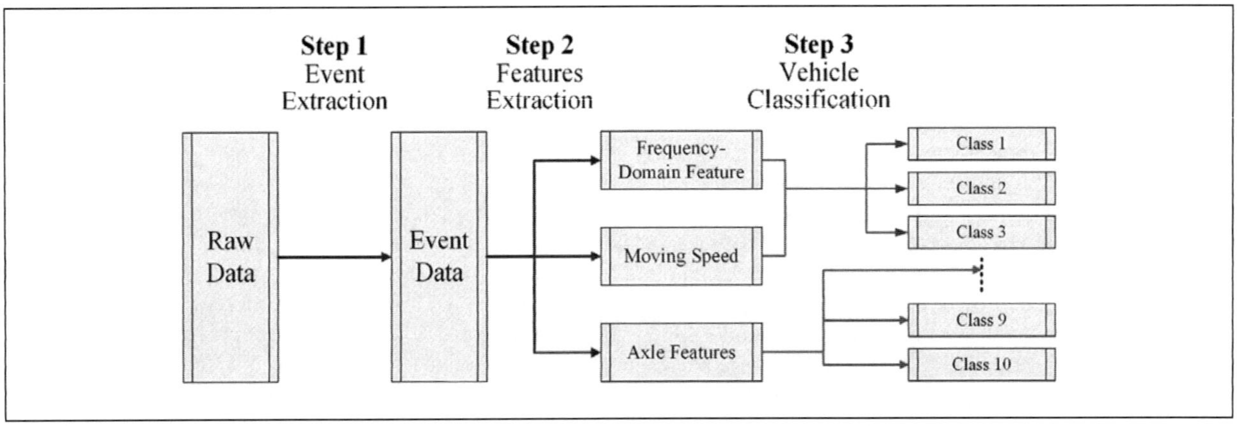

Figure 8. Classification procedures.

where
f_w = the weighted frequency,
$S(f)$ = the amplitude-frequency curve,
f_h = high limiting frequency,
f_l = low limiting frequency.

Based on J. Hajek's study (*15*), the vehicle-induced ground vibration comprises four components: (1) fundamental frequency (1~2 Hz), induced by suspended mass; (2) unsuspended mass-induced vibration (10~15 Hz); (3) pavement unevenness-caused vibration (0~40 Hz, related to moving speed); and (4) free vibration of pavement structure (usually over 20 Hz, related to the structural properties). Considering the high-frequency components (over 40 Hz, mainly concentrate in the fourth components), we selected the first three components to calculate the weighted frequency, and the f_h and f_l were set to be 0 Hz and 40 Hz, respectively.

Vehicle Classification and Validation

Classification Procedures

As shown in Figure 8, there are three steps in the designed vehicle classifier. Step 1 and Step 2 are event extraction and features extraction which are described in Section 3. The third step is vehicle classification based on multiple features. Heavy vehicles (multi-axle trucks and trailers, Class 4~10) are classified based on the axle configurations (number of axles and axle spacing). For the vehicles that possess similar axle features (vans, two-axle buses, and two-axle truck), a multi-parameter classifier is designed to identify those types of vehicles, the frequency-domain features and moving speeds are introduced.

Axle-Based Classification and Validation

In Step 3, the types of multi-axle vehicles (Class 6~10) are classified based on axle features. Firstly, multi-axle vehicles are identified if the number of axles is greater than three. Then, the specific types of the vehicles are identified based on their axle configurations. The classification results are provided in Table 2.

It can be seen that the overall accuracy is 89.36%, indicating that this prototype system shows a good performance in the classification of heavy vehicles. Obviously, the classification of single-unit trucks (class 4~6) and four-axle trailers (class 7) is more accurate than five-axle trailers (class 8) and six-axle trailers (Class 10), especially for those vehicles at high speed. Additionally, it is also noted that five five-axle trailers were misidentified as four-axle trailers, because the corresponding tridem-axles were identified as tandem-axles. The reason is that the axles in tridem-axle, compared with tandem-axle, have more effect on each other as there is one more axle, causing difficulties in identifying the middle axle of tridem-axle.

Two-Axle Vehicles Classification

As previously described, multi-axle vehicles are classified based on their axle configurations. However, the two-axle vehicles (class 1~3) cannot be identified because they have similar axle configurations. The vehicles of the three classes have the same numbers of axles, and there are overlaps between their axle spacing ranges (passenger car/van: 1.87~3.85 m; two-axle bus: 3.5~6.1m; two-axle truck: 3.3~6.2 m).

In this classifier, a SVM (Support Vector Machine) was utilized for multidimensional classification, as it is appropriate for small-sample clustering. Three features were selected for classification including axle spacing, weighted frequency (f_w), and moving speed. The 381 records of two-axle vehicles are divided into two sets: a training set (60 samples, including 20 passenger cars, 20 two-axle trucks, and 20 two-axle buses) and a testing set (321 samples). A radial basis function (RBF) is utilized as the kernel function:

$$k(x_i, x_j) = \exp\left(-\frac{\|x_i - x_j\|^2}{2\sigma^2}\right) \tag{18}$$

By using Matlab 2014b and LibSVM 3.2 (*16*), the classifier is designed based on the training set. The performance of the classifier is tested and validated through the testing set. The testing results are listed in Table 3.

Although the overall classification accuracy was 89.41%, the classification performance of two-axle trucks

Table 2. Classification Performance of Multi-Axle Vehicles

Class	Description	Actual number	Classified	Successful classified	Accuracy
4	Three-axle truck	10	10	10	100.00%
5	Three-axle truck	1	1	1	100.00%
6	Four-axle truck	10	10	10	100.00%
7	Four-axle trailer	50	51	46	92.00%
8	Five-axle trailer	19	14	14	73.68%
9	Five-axle trailer	1	1	1	100.00%
10	Six-axle trailer	3	2	2	66.67%
Overall		94	89	84	89.36%

Table 3. Classification Performance of Two-Axle Vehicles

	Class	Description	Actual number	Classified	Successful classified	Accuracy
Training set	1	Passenger car/van	20	22	20	100%
	2	Two-axle bus	20	21	20	100%
	3	Two-axle truck	20	17	17	85%
Testing set	1	Passenger car/van	214	230	206	92.38%
	2	Two-axle bus	12	12	11	91.67%
	3	Two-axle truck	95	79	70	73.68%
Overall			321	321	287	89.41%

(class 5) was unsatisfactory (73.68%). In the testing set, it is noted that over 20 two-axle trucks were misidentified as passenger cars/vans. The reason is that the frequency spectra of trucks are strongly related to the loading conditions. If the weight of a truck is sufficiently light and the axle spacing is sufficiently small, the induced vibration signal will behave more like a car than a truck.

Conclusion

In conclusion, this paper presents a vibration-based vehicle classification system using DOVS, and a prototype system was successfully installed and tested. The system utilizes in-pavement sensing fiber to monitor the traffic-induced vibrations with an interrogator to record the signals. A special layout is designed to improve the spatial localization accuracy.

A prototype system was implemented and tested in Shanghai. Using precast concrete pavement technology, the system was successfully installed on a concrete road. An EMD method, short-time energy analysis and STFT are employed for feature extraction. Based on short-time energy analysis, the algorithms that estimate axle configurations are proposed. The frequency-domain feature extraction is conducted via STFT.

Vibration signals induced by over 400 vehicles were collected using the prototype system, and then the classification performance was evaluated. Although results show a good classification accuracy of heavy vehicles, the identification accuracy of tridem axles is lower. For two-axle vehicles, the classification performance is good on passenger cars and buses, while the accuracy for two-axle truck classification is lower.

Future improvements to this system and classification method will be made through system installation and the multidimensional classification method. A more reliable method is needed to accelerate the system installation and prevent the system from being damaged. In addition, in the multidimensional classification, more features in the frequency domain will be considered to improve the classification accuracy, and more data is needed to validate classification performance.

Acknowledgments

This work was supported by the National Nature Science Foundation of China (Grant NO: 51778480) and China Scholarship Council.

Author Contributions

The authors confirm contribution to the paper as follows: study conception and design: Hongduo Zhao, Difei Wu; data analysis and draft manuscript preparation: Difei Wu, Mengyuan Zeng; field tests and data collection: Difei Wu, Mengyuan Zeng and Sheng Zhong. All authors reviewed the results and approved the final version of the manuscript.

References

1. Lee, H., and B. Coifman. Side-Fire Lidar-Based Vehicle Classification. *Transportation Research Record: Journal of the Transportation Research Board*, 2012. 2308: 173–183.
2. Yao, Z., H. Wei, Z. Li, and J. Corey. Fuzzy C -Means Image Segmentation Approach for Axle-Based Vehicle Classification. *Transportation Research Record: Journal of the Transportation Research Board*, 2016. 2595: 68–77.
3. Mimbela, L. E., and L. A. Klein. *Summary of Vehicle Detection and Surveillance Technologies Used in Intelligent Transportation Systems*. U.S. Federal Highway Administration, Intelligent Transportation System, 2007.
4. Ma, W., D. Xing, A. McKee, R. Bajwa, C. Flores, B. Fuller, and P. Varaiya. A Wireless Accelerometer-Based Automatic Vehicle Classification Prototype System. *IEEE Transactions on Intelligent Transportation Systems*, Vol. 15, No. 1, 2017, pp. 104–111.
5. Coifman, B., and S. Kim. Speed Estimation and Length Based Vehicle Classification from Freeway Single-Loop Detectors. *Transportation Research Part C*, Vol. 17, 2009, pp. 349–364.
6. He, Y., Y. Du, L. Sun, and Y. Wang. Improved Waveform-Feature-Based Vehicle Classification Using a Single-Point Magnetic Sensor. *Journal of Advanced Transportation*, Vol. 49, No. 5, 2015, pp. 663–682.
7. Kim, S., and B. Coifman. Evaluation of Axle-Based and Length-Based Vehicle Classification Stations. *Transportation Research Record: Journal of the Transportation Research Board*, 2013. 2339: 1–12.
8. Bajwa, R., R. Rajagopal, P. Varaiya, R. Kavaler, and N. Street. In-Pavement Wireless Sensor Network for Vehicle Classification Categories and Subject Descriptors. *Proc., 10th International Conference on Information Processing in Sensor Networks (IPSN)*, IEEE, Chicago, Ill., 2011.
9. Yeum, C. M., S. J. Dyke, and R. E. B. Rovira. Acceleration-Based Automated Vehicle Classification on Mobile Bridges. *Computer-Aided Civil and Infrastructure Engineering*, Vol. 31, 2016, pp. 813–825.
10. Stocker, M., P. Silvonen, M. Rönkkö, and M. Kolehmainen. Detection and Classification of Vehicles by Measurement of Road-Pavement Vibration and by Means of Supervised Machine Learning Detection and Classification of Vehicles by Measurement of Road-Pavement Vibration and by Means of Supervised Machine. *Journal of Intelligent Transportation Systems*, Vol. 20, No. 2, 2016, pp. 125–137.
11. Bao, X., and L. Chen. Recent Progress in Distributed Fiber Optic Sensors. *Sensors*, Vol. 12, No. 7, 2012, pp. 8601–8639.
12. Taylor, H. F., and C. E. Lee. *Apparatus and Method for Fiber Optic Intrusion Sensing*. U.S. Patent and Trademark Office, Washington, D.C., 1993.
13. Shatalin, S. V., V. N. Treschikov, and A. J. Rogers. Interferometric Optical Time-Domain Reflectometry for Distributed Optical-fiber Sensing. *Applied Optics*, Vol. 37, No. 24, 1998, pp. 5600–5604.

14. Huang, N. E., Z. Shen, S. R. Long, M. C. Wu, H. H. Shih, Q. Zheng, N. C. Yen, C. C. Tung, and H. H. Liu. The Empirical Mode Decomposition and the Hilbert Spectrum for Nonlinear and Non-Stationary Time Series Analysis. *Proceedings Mathematical Physical and Engineering Sciences*, Vol. 454, No. 1971, 1998, pp. 903–995.
15. Hajek, J., C. Blaney, and D. Hein. Mitigation of Highway Traffic-Induced Vibration. *Proc., Annual Conference of the Transportation Association of Canada*, Charlottetown, Canada, 2006.
16. Chang, C. -C., and C. -J. Lin. *LIBSVM – A Library for Support Vector Machines*. https://www.csie.ntu.edu.tw/~cjlin/libsvm/. 2016. Retrieved July 20, 2017.

The Standing Committee on Highway Traffic Monitoring (ABJ35) peer-reviewed this paper (18-00201).

Article

TRR
JOURNAL OF THE TRANSPORTATION RESEARCH BOARD

Transportation Research Record
2018, Vol. 2672(43) 24–33

DOI: 10.1177/0361198118774194
journals.sagepub.com/home/trr
SAGE

How Accurately Can We Measure from Video? Practical Considerations and Enhancements of the Camera Calibration Procedure

Aliaksei Laureshyn[1] and Mikael Nilsson[2]

Abstract

The accuracy of position measurements from videos depends greatly on the quality of the camera calibration model parameters. This paper investigates how such factors as camera height and the selection of calibration points affect the quality of the final calibration model. A series of controlled experiments were performed in traffic or similar-to-traffic environments, in which the accuracy of measurements from videos was compared with measurements taken with other tools. To enhance the calibration process, a multi-camera approach is suggested that utilizes the information about "common points" – points seen on several cameras but with unknown world coordinates. The performed tests showed that calibration quality can greatly benefit from this approach. The paper is addressed primarily to traffic researchers developing their own video-based tools for road user observations.

The video recording and extraction of microscopic data, such as road user position and speed, from videos is a commonly used method in many road traffic studies. Examples of such applications include surrogate safety analysis (*1–5*), behavioral analysis (*6*, *7*), and the calibration of microscopic traffic models (*8–10*). A crucial step before objective measurements from a video can be taken is camera calibration, that is, the definition of a model that relates the position of a point **A** in world coordinates (x_w, y_w, z_w) to its representation **a** (x_i, y_i) in the camera image (Figure 1).

Several techniques for camera calibration have been suggested, the most common being Heikkilä and Silven, Tsai, and Zhang (*11–13*). This study uses the Tsai calibration model, which incorporates linear equations based on the radial alignment constraint and a second-order radial term that handles lens distortions (the exact mathematical description can be found elsewhere, e.g., Tsai [*12*]).

One important mathematical property of the calibration model deserves a mention here. Transforming three-dimensional (3D) world coordinates to an image can be done unambiguously, but this is not the case for transforming in the opposite direction (from an image to the world). An image point corresponds to a series of points in the 3D space or, more specifically, all the points along a ray (see the dashed line in Figure 1). Therefore, it is necessary to have additional information (usually, the point's *z*-coordinate) to define a single corresponding point in the real world. This problem is often tackled using an approximation that states the part of the road surface visible in an image is a perfect plane, that is, the *z*-coordinate of the road user footprint is always zero (we will refer to this later as the "zero-plane").

The parameters of the calibration model with a fixed mounted camera are usually estimated based on a set of points with a known position in an image and known world coordinates. Once again, we omit the heavy mathematical description of the process, which can be found, for example, in Tsai (*12*).

Video-based tools available on the market usually come with a list of recommendations on how the cameras should be installed (*14*). However, a traffic researcher willing to develop their own tools with functionality beyond what the "standard sensors" have to offer, usually has to find his/her own way by trial and error, which might often be costly in time, energy, and amount of frustration. The purpose of this paper is to provide practical hints and a better understanding about which factors affect the quality of the camera calibration model and consequently the measurements taken from a video and the extent of these effects. Several real-world experiments have been conducted testing the different factors, such as camera perspective, location of the points used for calibration, and so forth. A multi-camera approach is also suggested to enhance the quality of the calibration model.

In this study, we use used a software tool called T-Calibration (*15*), which is available as a free download.

[1]Transport & Roads, Department of Technology & Society, Faculty of Engineering, LTH, Lund University, Lund, Sweden
[2]Centre of Mathematical Sciences, Faculty of Engineering, LTH, Lund University, Lund, Sweden

Corresponding Author:
Address correspondence to Aliaksei Laureshyn:
aliaksei.laureshyn@tft.lth.se

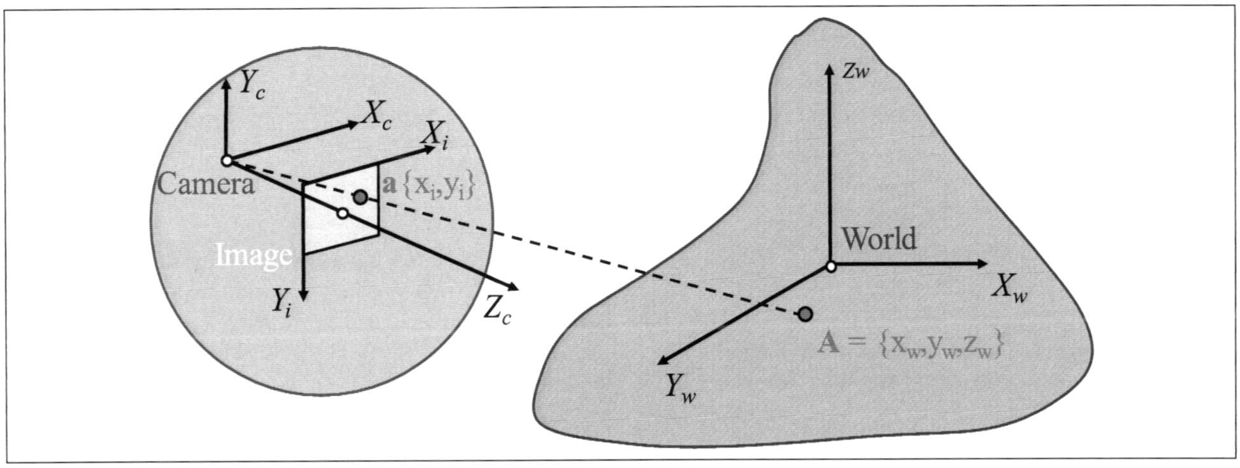

Figure 1. Calibration model principle.

Figure 2. The "size" of one camera image pixel in real-world distance units.

The software is based on Tsai's model description (*12*) except for the module on model parameters' initialization, which utilizes the code of Reg Willson (*16*).

What Affects the Accuracy of Measurements from a Video?

A long list of factors can affect the accuracy of measurements taken from a video. The most important are:

- Camera height and perspective,
- Camera resolution,
- The distance from the camera to the studied object (see Figure 2),
- The quality of the calibration model.

An objective measure of the calibration model quality is the discrepancy between the actual position of a calibration point in the image and the projection to the image of its world coordinates (further noted as *e*, measured in pixels) or, alternatively, the actual position of a calibration point in the world and its projection from the image (further noted as *E*, measured in meters) – see Figure 3. As the optimization criteria of the calibration model parameters, minimization of one of the following parameters can be used:

- $\bar{E}$, m – average of the discrepancies between the points in world coordinates and their projections from the camera view,
- E_{max}, m – maximum discrepancy value in world coordinates,
- $\bar{e}$, pxl – average of the discrepancies between the points and their projections from the world in the camera image,
- e_{max}, pxl – maximum discrepancy value in the camera image.

However, the final judgment of the obtained calibration model is as much an art as it is a precise science. Low values

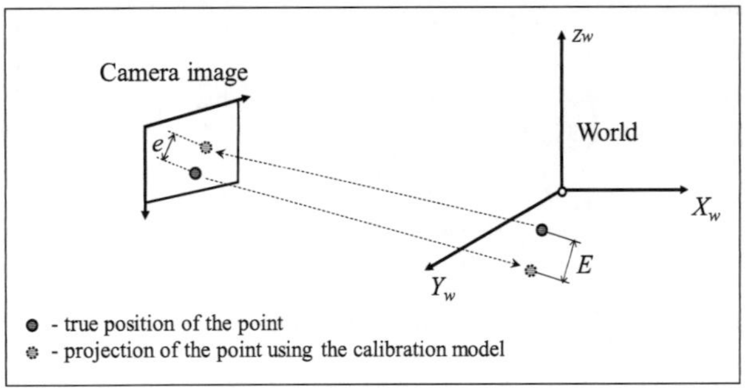

Figure 3. Discrepancies between the true and "projected" positions of a calibration point due to imperfections of the calibration model.

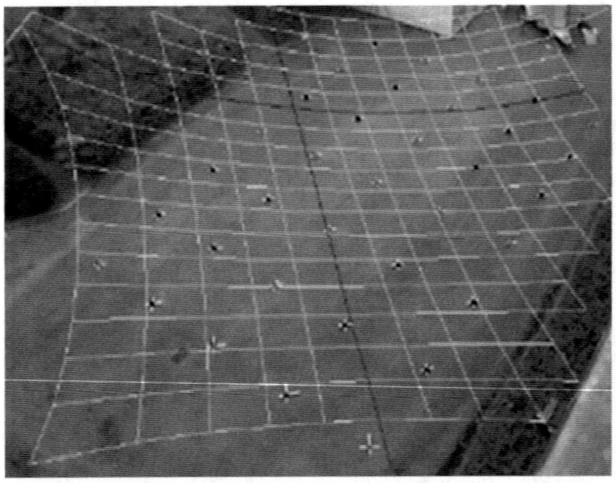

Figure 4. The grid placed on the road surface suggests the poor quality of the calibration model, even though the error metrics on the calibration points are low.

for E and e do not necessarily guarantee an accurate calibration model. In practice, the number of available calibration points is usually limited, and these points do not necessarily cover the entire scene as evenly as desired. In this case, the optimization may "force" the model to fit the available points well, but for the areas with no points, the calibration quality will remain uncertain.

At least some initial clues regarding the general model fit can be obtained from a visual check. This can be done by, for example, placing a grid of regularly positioned points in the road plane and projecting them to the camera view. Strange curvatures and distortions (Figure 4) would indicate that a further enhancement of the calibration model is necessary.

Dataset I

Data Collection Set-Up

A grid of 35 points (cell size 2 m) was marked on a flat, traffic-free space surrounded by building blocks (Figure 5). Cameras 1 and 2 were installed almost exactly above each other on the outdoor emergency ladder (heights 5.5 m and 8 m respectively) and cameras 3 and 4 on the roof (height 11 m). The points' positions were measured accurately with a theodolite (measurement error <1 cm). The height difference between the individual points did not exceed 1 to 2 cm, and, despite being considered in all calculations, it did not produce any noticeable effect.

This set-up is much more accurate compared with what is considered "normal" accuracy in traffic studies. In a sense, it is a controlled simulation, close enough to real road environments, but it allows for testing of the effects of different factors, such as the selection of calibration points, camera angle, and introduced errors in calibration point coordinates.

Experiment I

The main objective of this experiment was to investigate how the camera angle and the selection of calibration points affect the final calibration model quality. Views from cameras 1, 2, and 3 were used, as their views of the calibration points were highly similar (though "mirrored" for Camera 3), except for the camera installation heights. The following scenarios were tested (Figure 6):

- ***S1***: all the available points used for calibration;
- ***S2***: few points evenly distributed over the outer border of the studied area and no points in the middle; this is a typical situation for an urban intersection where many calibration points can be found at zebra crossings around the intersection, but there are no points in its central part;
- ***S3***: few points concentrated in the corner close to the camera;
- ***S4***: few points in the corner far away from the camera.

The results are presented in Figure 6 and Table 1. Scenario S1 is the "perfect case," in which many calibration points are evenly distributed over the entire image. All three camera installations provide excellent accuracy, even though it can be noted that, in full accordance with the basic rules of geometry, real-world errors slightly increase as the camera height lowers.

Scenario S2, despite the increase in errors, still appears quite usable, as the points, though not as frequent as in S1, still cover a large area of the image. Adversely, scenarios S3 and S4 clearly demonstrate how the accuracy deteriorates as the measurements are attempted in the image areas that do not have any calibration points in the vicinity. A bit surprisingly (and against what theory would suggest), the increased camera height does result in systematically improved accuracy from Camera 1 to Camera 3 within these scenarios. It appears the location of the calibration points has a much stronger effect on the overall accuracy compared with what can be gained from the increased camera height.

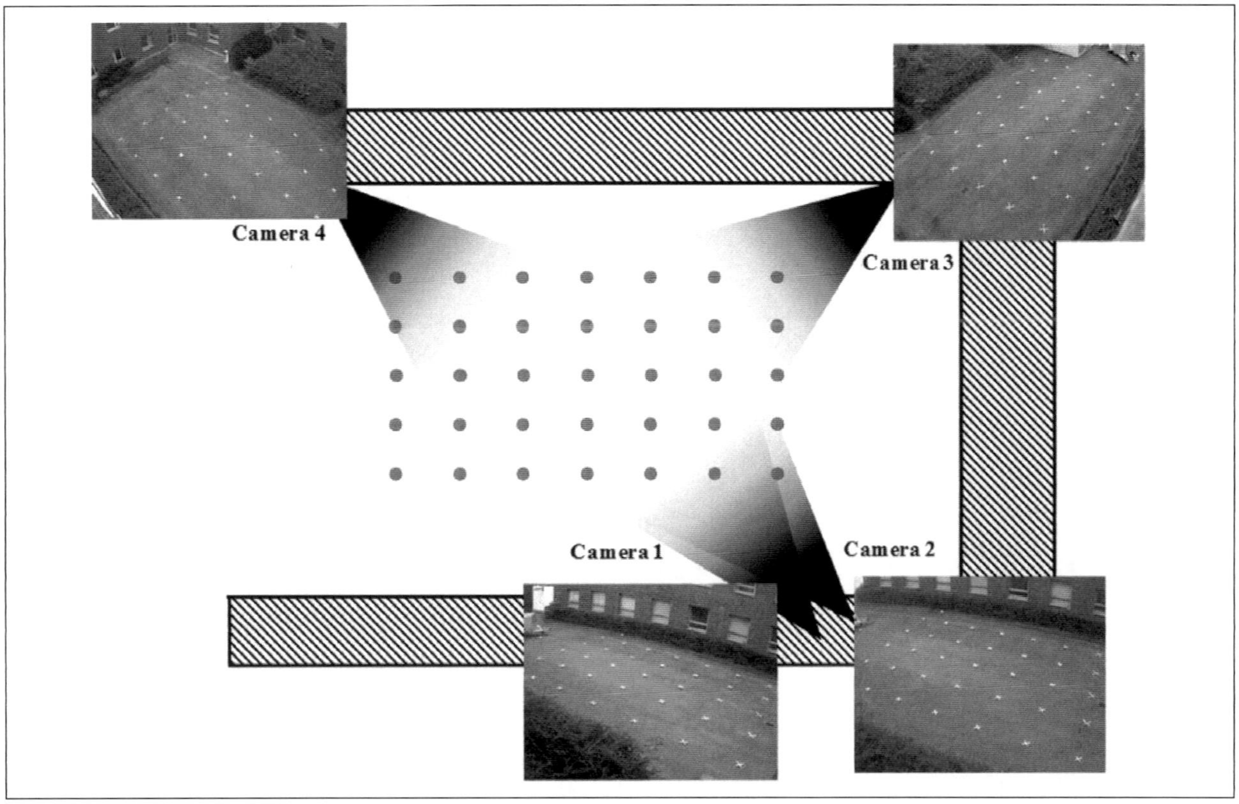

Figure 5. Camera set-up for Dataset I.

Experiment 2

This experiment investigates how the accuracy of the world coordinates for the calibration points affects the calibration model's quality. It is common that instead of measuring the points with a theodolite or similar high-accuracy instrument in the field, their positions are extracted from satellite photos provided by such services as Google Earth or Bing Maps. Such photos are often blurry and outdated, and even though they provide a good general overview and indicate approximately where the point should be, the actual position measured is not quite accurate.

In this test, we used the points from Scenario S2 in the previous experiment, as it is the most typical situation in the case of traffic videos. The actual world coordinates of the calibration points were modified with added errors *dx* and *dy*, generated randomly from a normal distribution with $\mu = 0$ and chosen σ. The calibration procedure was repeated 100 times with newly generated *dx* and *dy* values. The final distributions of the calibration errors are presented in Figure 7.

If the random variables that characterize the population are normally distributed, then there is approximately a 95% probability that the sample mean is within ±2 standard deviations of the population mean (i.e., it can be roughly assumed that $\sigma = 10$ cm represents the accuracy of the point measurement ±20 cm). It can be seen from Figure 7 that even with a small σ, there are some cases with an E_{max} value up to 80 cm, which is clearly unusable. This indicates how important it is to ensure that the obtained calibration model really makes sense, for example, using a visual check (Figure 4). "Guessing" the position of the points, which are not really seen in the satellite image, is probably not advisable either.

Experiment 3

In this test, we examined whether the use of a multi-camera set-up could improve the calibration quality of each individual camera. It is common that the potential calibration points clearly detectable in the satellite image are not visible in the camera view and vice versa. However, if several camera views are available, points not seen on one camera might be seen on another, and so forth. Moreover, there are usually many "common" points found in all camera views, but not in the satellite image. Examples of such points can include water traces on asphalt, shadows, or the wheel position of a stopped vehicle (given that the camera images are taken at approximately the same time).

The following algorithm for multi-camera calibration is suggested.

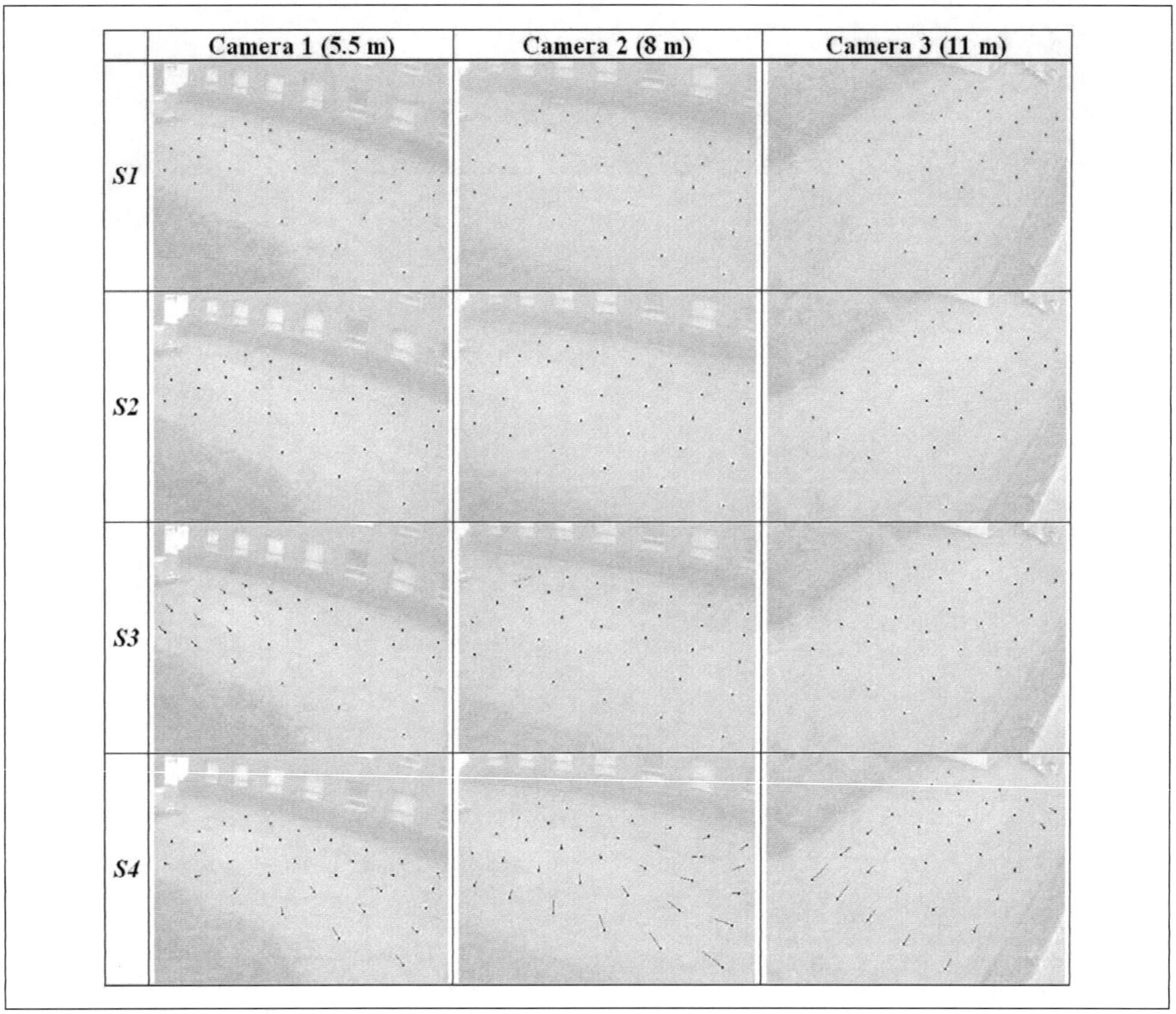

Figure 6. Calibration point selection for Experiment 1. Points used for calibration are marked red. Error vectors connect the points in the image with the projection of the corresponding point from the world coordinates. Note how error vectors increase with distance from the calibration points in scenarios S3 and S4.

1. Each camera is calibrated independently based on "few" points that can be identified in both the satellite image and the camera view.
2. The "common" points are projected to the world coordinates using the available calibration models.
3. Because the world coordinates of a point suggested by each camera differ, they are "weighed" into a single value in the simplest case by taking an average.
4. Each camera is again calibrated, now using an extended set of points with world coordinates.
5. Steps 3 to 5 are repeated until further iterations do not yield any improvements.

To test this approach, we used the views from cameras 1 and 4. For the initial calibration, a highly limited and unfavorable set of points was selected (Figure 8). These were complemented by 10 common points marked in each camera but without the real-world coordinates. After approximately 10 iterations, no visible changes were produced. The results of the initial individual calibration and final multi-camera calibration are presented in Table 2.

The use of additional calibration points, despite their estimated and not perfectly accurate real-world position, greatly improved the calibration accuracy, particularly for Camera 1, in which both the average and maximum errors greatly decreased. For Camera 4, the average error did not change, but the maximum error decreased, indicating the calibration became more consistent over the entire image.

Table 1. Calibration Quality Parameters for Experiment 1

		Camera 1 5.5 m	Camera 2 8 m	Camera 3 11 m
S1	Points	34	34	35
	$\bar{E}$, cm	2	1	1
	E_{max}, cm	9	5	3
	$\bar{e}$, pxl	0	0	0
	e_{max}, pxl	1	1	1
S2	Points	6	6	6
	$\bar{E}$, cm	3 (2)	2 (1)	3 (1)
	E_{max}, cm	10 (10)	5 (4)	8 (3)
	$\bar{e}$, pxl	1 (1)	1 (0)	1 (0)
	e_{max}, pxl	3 (1)	2 (2)	2 (1)
S3	Points	5	5	5
	$\bar{E}$, cm	25 (1)	8 (0)	19 (1)
	E_{max}, cm	93 (5)	35 (1)	73 (2)
	$\bar{e}$, pxl	5 (0)	2 (0)	4 (0)
	e_{max}, pxl	19 (2)	5 (0)	9 (1)
S4	Points	5	5	5
	$\bar{E}$, cm	24 (2)	42 (4)	31 (0)
	E_{max}, cm	87 (8)	92 (13)	149 (1)
	$\bar{e}$, pxl	8 (0)	15 (1)	10 (0)
	e_{max}, pxl	28 (1)	54 (2)	47 (0)

Note: For the discrepancies *E* and e, the main values in the table are calculated for all the calibration points. Values in brackets are calculated only for the points actually used for calibration.

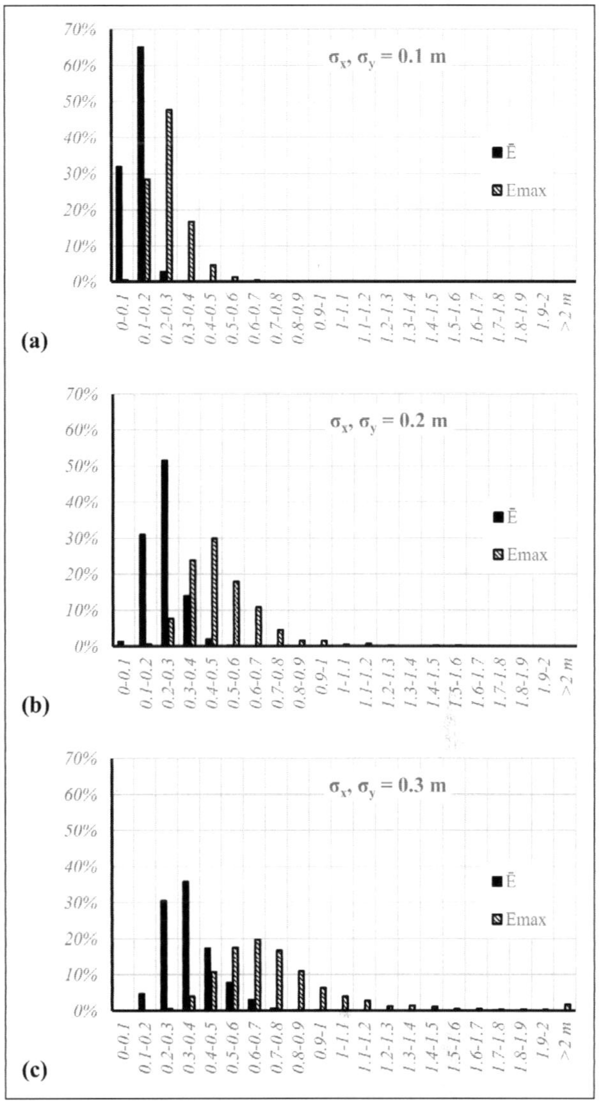

Figure 7. Distribution of the calibration $\bar{E}$ and E_{max} for simulated "inaccurate" world coordinates of the calibration points.

Dataset II

Data Collection Set-Up

The second location was a real traffic intersection (Figure 9). It was filmed with a camera installed on the top of a water tower. An additional photo image was taken from a building on the opposite side of the intersection. The calibration points were limited to the actual landmarks, such as sharp corners of the road markings. The calibration point position was measured with a high-precision GNSS receiver (Leica GX1230 GG, http://www.leica-geosystems.com/), providing accuracy for static points of 1.5 to 2 cm. Most of the measured points could also be found in a satellite image from Googl Maps.

A car, equipped with the same type of GNSS receiver, drove through the intersection several times, performing different maneuvers. The position was saved with a frequency of 5 Hz; however, the accuracy of the measurement was affected by the movement and short time for each measurement, and this varied from 5 to 70 cm (RMS). Only measurements with an accuracy below 30 cm were used.

Experiment 4

In this experiment, we tested how well the trajectories of road users extracted from video (with different calibration models) fit the trajectories obtained from the Global Navigation Satellite Systems (GNSS) tool. To extract the trajectory of a road user, a software tool, T-Analyst, was used (*15*). This tool allows the user to browse a video frame by frame and to set a pre-defined wire-frame model so that it fits the road user in the image (Figure 10). The parameters of the wire-frame model were based exactly on the measurements of the car used in the experiment.

The following calibration scenarios were tested.

- All available calibration points with measured world coordinates used (Figure 11);
- Few poorly located points with world coordinates taken from a Google Maps satellite image (Figure 12);
- Multi-camera set-up calibration according to the algorithm described in Experiment 3. The initial calibration

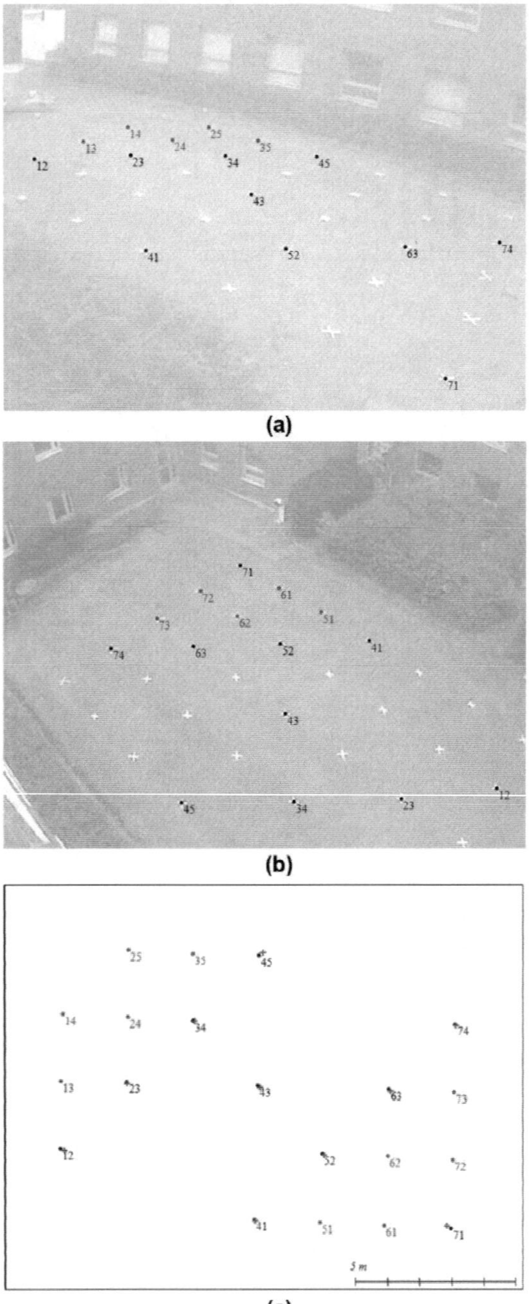

Figure 8. Multi-camera calibration: (*a*) Camera 1 view, (*b*) Camera 4 view, and (*c*) mapping of the calibration points in the real world. In red, the original points with known real-world positions; in black, the "common points" and their true real-world positions; red cross, the estimated positions of the common points in the real world.

of both cameras was undertaken with few and unfavorably located points and the world coordinates taken from a satellite image (Figure 13).

Table 2. Calibration Quality with Single- and Multi-Camera Approaches

	Individual calibrations		Multi-set-up calibrations	
	Camera 1	Camera 4	Camera 1	Camera 4
$\bar{E}$, cm	20	6	9	7
E_{max}, cm	66	25	24	14
$\bar{e}$, pxl	8	2	3	2
e_{max}, pxl	41	9	9	6

Not surprisingly, calibration based on few points (second scenario) failed completely. However, when the same set of points was used together with another camera view in a multi-camera calibration, highly satisfactory results were obtained. In fact, it appears that the multi-camera calibration provides a better fit with the GNSS-obtained trajectories compared with the calibration based on accurately measured points in the first scenario. An explanation might be that the availability of the height data for points in the first scenario allows for an estimation of the correct horizontal plane on which road users can move, even though the actual road surface was not horizontal. The point coordinates obtained from the satellite image lack height data; thus, the estimated zero-plane is not horizontal, and it is more closely adjusted to the actual road surface.

Generally, the assumption that road users are moving on the zero-plane might result in inaccuracies if the studied area is not flat (for example, if one of the intersection legs had a more acute incline compared with the rest of the intersection). This becomes even more probable if a larger area is studied (e.g., covered with several cameras). To compensate, it might be necessary to create a model of the road surface (for example, by interpolating the heights of the measured points); instead of always assuming that road users travel on the zero-plane, the actual height calculated by the surface model could be used.

Conclusions

Carefully choosing calibration points appears highly crucial to the quality of the final calibration model. Ideally, these points should cover evenly the entire camera image, but their locations around the studied area (e.g., zebra markings around an intersection) might still be a reasonable compromise.

Camera height affects the accuracy of the measurements, but to a lesser extent than calibration point selection.

Inaccurate world coordinates of the calibration points (e.g., taken from a blurry satellite image) result easily in an inadequate calibration model. Additional visual control of the obtained calibration (e.g., by drawing a grid on the zero-plane) is highly recommended.

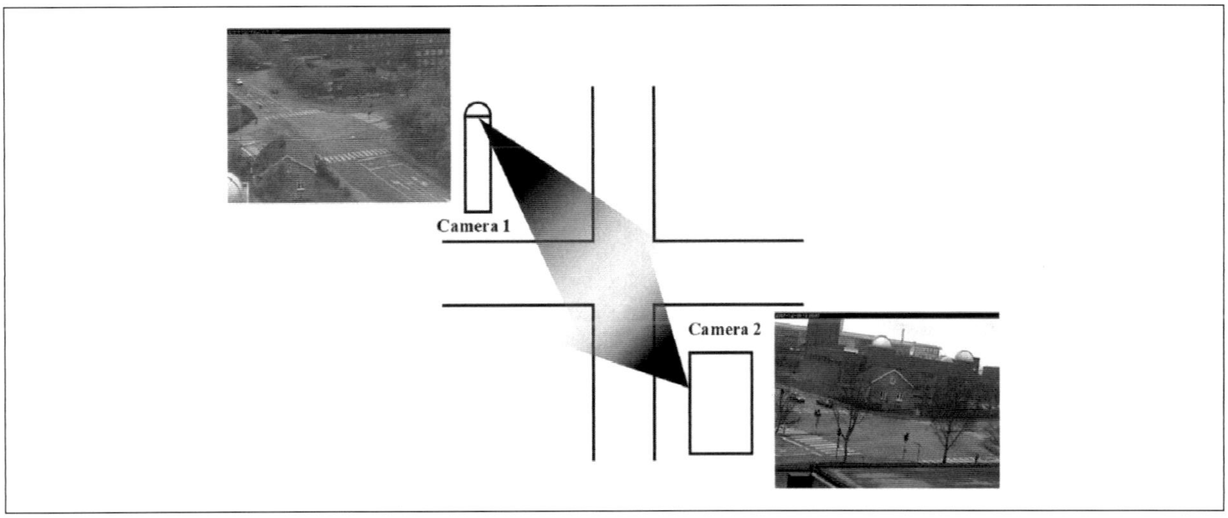

Figure 9. Camera set-up for Dataset II.

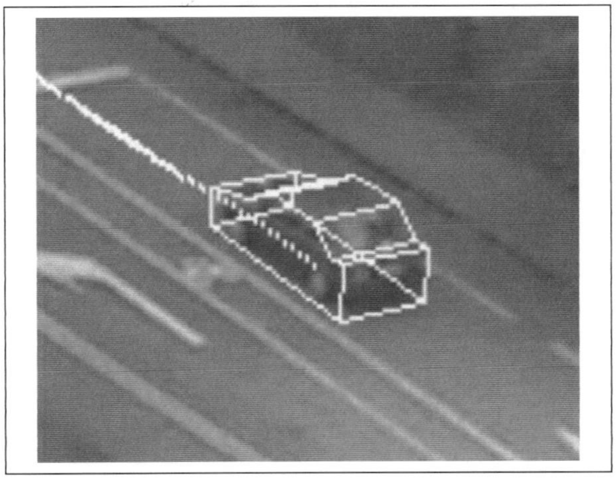

Figure 10. Wire-frame model.

The use of several camera views with common points may greatly improve the calibration of each single camera. The additional camera views do not necessarily require another video camera installation, they can simply be taken with a mobile phone or camera, thus minimizing the additional work required.

The introduction of a road surface height model might be an additional step toward improving the accuracy of measurements from videos. This might be particularly important in the case of a large study area in which significant height variations are possible; thus, the assumption that road users move on the zero-plane might be insufficient.

In this study, we have not investigated the temporal aspect of the measurements from videos (for example, the accuracy of speed and acceleration calculated based on the positions measured from a video). This aspect should be addressed in future research.

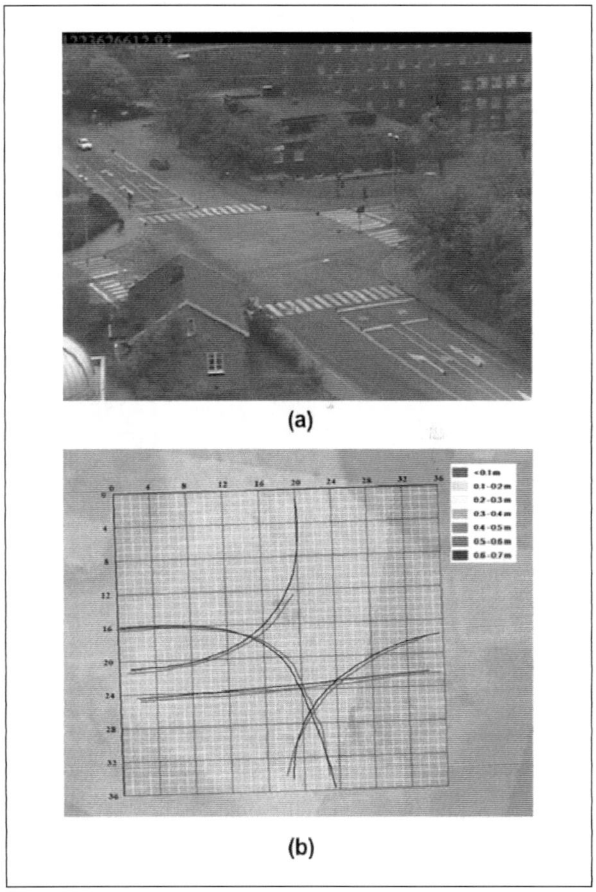

Figure 11. Calibration based on measured 3D points: (*a*) calibration points and (*b*) trajectories obtained from GNSS (black) and video (red). The colors in the background show the "price" of an image pixel error in world distance units; it should be considered that as the price increases, a lower accuracy of the video-based trajectory points should be expected.

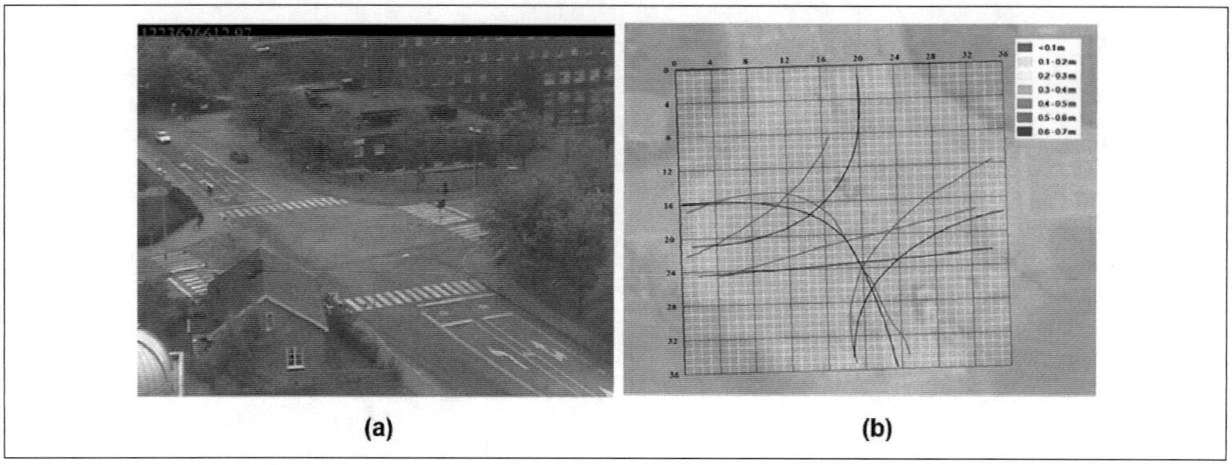

Figure 12. Calibration based on few points and world coordinates obtained from a satellite image: (*a*) calibration points and (*b*) trajectories from GNSS (black) and video (red).

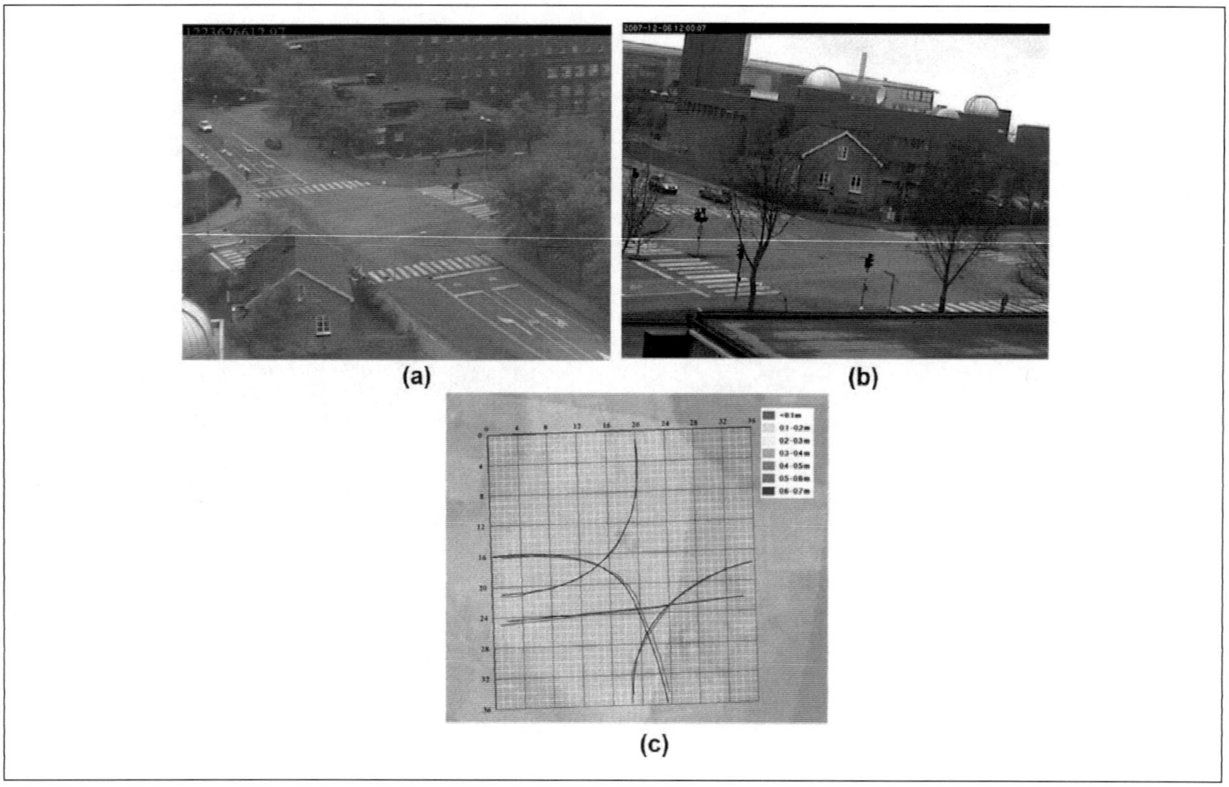

Figure 13. Multi-camera calibration: (*a*) and (*b*) points used for initial calibration (red) and "common points" without known world coordinates (green); (*c*) trajectories obtained from GNSS (black) and video (red).

Acknowledgments

This work has received funding from the European Union's Horizon 2020 research and innovation program under grant agreement No. 635895. Some data were collected in a project financed by Vinnova, Sweden's innovation agency.

Author Contributions

The authors confirm contribution to the paper as follows: study conception and design: A.L., M.N.; data collection: A.L.; analysis and interpretation of results: A.L.; draft manuscript preparation: A.L., M.N. All authors reviewed the results and approved the final version of the manuscript.

References

1. Laureshyn, A., M. de Goede, N. Saunier, and A. Fyhri. Cross-comparison of Three Surrogate Safety Methods to Diagnose Cyclist Safety Problems at Intersections in Norway. *Accident Analysis & Prevention*, Vol. 105, 2017, pp. 11–20. http://dx.doi.org/10.1016/j.aap.2016.04.035.
2. Saunier, N., T. Sayed, and K. Ismail. Large Scale Automated Analysis of Vehicle Interactions and Collisions. *Transportation Research Record: Journal of the Transportation Research Board*, 2010. 2147: 42–50.
3. Madsen, T., and H. Lahrmann. Comparison of five bicycle facility designs in signalized intersections using traffic conflict studies. *Transportation Research Part F: Traffic Psychology and Behaviour*, No. 46 (Part B), 2017, pp. 438-450. https://doi.org/10.1016/j.trf.2016.05.008.
4. van der Horst, A. R. A., M. C. Thierry, J. M. Vet, and A. F. Rahman. An Evaluation of Speed Management Measures in Bangladesh Based upon Alternative Accident Recording, Speed Measurements, and DOCTOR Traffic Conflict Observations. *Transportation Research Part F: Traffic Psychology and Behaviour*, Vol. 46, 2017, pp. 390–403. http://dx.doi.org/10.1016/j.trf.2016.05.006.
5. St-Aubin, P., L. Miranda-Moreno, and N. Saunier. A Surrogate Safety Analysis at Protected Freeway Ramps Using Cross-Sectional and Before-After Video Data. Presented at 91st Annual Meeting of the Transportation Research Board, Washington, D.C., 2012.
6. Polders, E., J. Cornu, T. De Ceunynck, S. Daniels, K. Brijs, T. Brijs, E. Hermans, and G. Wets. Drivers' Behavioral Responses to Combined Speed and Red Light Cameras. *Accident Analysis & Prevention*, Vol. 81, 2015, pp. 153–166. http://dx.doi.org/10.1016/j.aap.2015.05.006.
7. De Ceunynck, T., B. Dorleman, S. Daniels, A. Laureshyn, T. Brijs, E. Hermans, and G. Wets. Sharing is (s)caring? Interactions between Buses and Bicyclists on Bus Lanes Shared with Bicyclists. *Transportation Research Part F: Traffic Psychology and Behaviour*, Vol. 47, 2017, pp. 301–315. http://dx.doi.org/10.1016/j.trf.2016.09.028.
8. Archer, J. *Indicators for Traffic Safety Assessment and Prediction and Their Application in Micro-Simulation Modelling: A Study of Urban and Suburban Intersections*. Doctoral thesis. Department of Infrastructure, Royal Institute of Technology, Stockholm, 2005.
9. Huang, F., P. Liu, H. Yu, and W. Wang. Identifying if VISSIM Simulation Model and SSAM Provide Reasonable Estimates for Field Measured Traffic Conflicts at Signalized Intersections. *Accident Analysis & Prevention*, Vol. 50, 2013, pp. 1014–1024. http://dx.doi.org/10.1016/j.aap.2012.08.018.
10. Wu, J., E. Radwan, and H. Abou-Senna. Determine if VISSIM and SSAM could Estimate Pedestrian-Vehicle Conflicts at Signalized Intersections. *Journal of Transportation Safety & Security*, 2017 (online). https://doi.org/10.1080/19439962.2017.1333181.
11. Heikkilä, J., and O. Silven. A Four-Step Camera Calibration Procedure with Implicit Image Correction. *Proc., IEEE Conference on Computer Vision and Pattern Recognition, IEEE*, New York, 1997, pp. 1106–1112.
12. Tsai, R. Y. A Versatile Camera Calibration Technique for High-Accuracy 3D Machine Vision Metrology Using Off-the-Shelf TV Cameras and Lenses. *IEEE Journal of Robotics and Automation*, Vol. RA-3, No. 4, 1987, pp. 323–344. https://doi.org/10.1109/JRA.1987.1087109.
13. Zhang, Z. A Flexible New Technique for Camera Calibration. *IEEE Transactions on Pattern Analysis and Machine Intelligence*, Vol. 22, No. 11, 2000, pp. 1330–1334.
14Klein, L. A. *ITS Sensors and Architectures for Traffic Management and Connected Vehicles*. CRC Press, Taylor & Francis Group, Boca Raton, FL, 2018. https://doi.org/10.1201/9781315206905.
15. T-Analyst. Software for Semi-Automated Video Processing. www.tft.lth.se/software. Accessed April 4, 2017.
16. Willson, R. Tsai Code Implementation. http://www-2.cs.cmu.edu/afs/cs.cmu.edu/user/rgw/www/TsaiCode.html. Accessed January 15, 2014.

The Standing Committee on Highway Traffic Monitoring (ABJ35) peer-reviewed this paper (18-01963).

This publication reflects only the authors' views. The European Commission and Vinnova are not responsible for any use of the information it contains.

TRR
JOURNAL OF THE TRANSPORTATION RESEARCH BOARD

Article

Transportation Research Record
2018, Vol. 2672(43) 34–43

DOI: 10.1177/0361198118790619
journals.sagepub.com/home/trr
SAGE

Evaluating the Reliability, Coverage, and Added Value of Crowdsourced Traffic Incident Reports from Waze

Mostafa Amin-Naseri[1], Pranamesh Chakraborty[1], Anuj Sharma[1], Stephen B. Gilbert[1], and Mingyi Hong[2]

Abstract

Traffic managers strive to have the most accurate information on road conditions, normally by using sensors and cameras, to act effectively in response to incidents. The prevalence of crowdsourced traffic information that has become available to traffic managers brings hope and yet raises important questions about the proper strategy for allocating resources to monitoring methods. Although many researchers have indicated the potential value in crowdsourced data, it is crucial to quantitatively explore its validity and coverage as a new source of data. This research studied crowdsourced data from a smartphone navigation application called Waze to identify the characteristics of this social sensor and provide a comparison with some of the common sources of data in traffic management. Moreover, this work quantifies the potential additional coverage that Waze can provide to existing sources of the advanced traffic management system (ATMS). One year of Waze data was compared with the recorded incidents in the Iowa's ATMS in the same timeframe. Overall, the findings indicated that the crowdsourced data stream from Waze is an invaluable source of information for traffic monitoring with broad coverage (covering 43.2% of ATMS crash and congestion reports), timely reporting (on average 9.8 minutes earlier than a probe-based alternative), and reasonable geographic accuracy. Waze reports currently make significant contributions to incident detection and were found to have potential for further complementing the ATMS coverage of traffic conditions. In addition to these findings, the crowdsourced data evaluation procedure in this work provides researchers with a flexible framework for data evaluation.

Traffic managers aim for increased mobility and safety on the roads. Real-time information on road conditions is necessary for taking proper actions. However, relying on sensors and cameras for monitoring traffic conditions at all locations and times is neither possible nor economically justifiable (*1*). Moreover, many sensors detect incidents based on speed changes, whereas in less populated areas, a crash may present a high-risk zone for secondary crashes without an immediate significant speed drop. These circumstances point to the insufficiency of the existing means for full road condition monitoring.

Recent research has demonstrated the potential value in leveraging social media to detect traffic incidents (*2–5*). Thus, crowdsourced data have recently gained attention in traffic management. To this end, many cities and departments of transportation (DOTs) have incorporated data from a crowdsourced smartphone application called Waze into their advanced traffic management system (ATMS). Using crowdsourced data, however, raises several questions for traffic managers. In this research, a quantitative analysis is implemented to provide data-driven answers to some of the common concerns of traffic managers with regards to Waze data.

Iowa Department of Transportation (IDOT) has used Waze data as a source of incident detection since September 2015. One year of data (2016) was used to address questions in three primary areas.

a. How does Waze compare with existing sources?
   - Are Waze reports reliable?
   - What percentage of the current recorded incidents were detected by Waze?
   - How does Waze compare with other common sources of data collection in the ATMS?
b. What are the characteristics of Waze data?

[1]Iowa State University, Ames, IA
[2]University of Minnesota, Minneapolis, MN

Corresponding Author:
Address correspondence to Mostafa Amin-Naseri: aminnas@iastate.edu

- How does Waze coverage compare with other sources?
- How does Waze coverage vary by time and location?

c. What is the estimated potential additional coverage that Waze can provide to the ATMS?
- In the locations where ATMS is unable to verify Waze reports, can Waze be trusted?

This last question is a critical topic. In current ATMS settings, crowdsourced data needs validation by a second source before being trusted. This is not available in all locations and times, however. Thus, an estimation of the potential added coverage in Waze provides a ground for justifying allocating resources to developing methods that assess crowdsourced reports using historical data. One of the ultimate goals of studying crowdsourced data is to understand its characteristics profoundly enough to know when and where to rely on crowdsourced reports in locations where there are no other means for validation. Therefore, this work seeks answers to the above questions in the process of finding the response to question (c). Moreover, some of the main challenges in utilizing Waze data for traffic monitoring were identified and discussed for future work.

Background

Crowdsourced data and social media have been widely used in many areas. For instance, tweets have been used to detect earthquakes in real time (*6*) or predict influenza outbreaks (*7*, *8*). More closely related to traffic, the Twitter-based Event detection and Analysis System (TEDAS) has been proposed by Li and colleagues (*3*). Another work utilized Twitter to detect traffic incidents in real time (*4*). To increase the percentage of useful tweets, Gu et al. have implemented a method to extract geolocation from the text of traffic-related tweets (*2*). Furthermore, the validity of the traffic information acquired from social media was approved by comparing with the recorded traffic situation in London (*9*). These applications demonstrate the potential wealth of information in crowdsourced data. Regardless of how the data are collected, however, there are challenges in using crowdsourced data that require consideration.

Although crowdsourced data usually come at a relatively inexpensive price, there are challenges in understanding and interpreting this type of data. The crowdsourced data are reported by users who might be slightly inaccurate in time or location. For users traveling on the roads at the speed of 60 miles per hour, 30 seconds' delay in reporting an incident is a 0.5-mile distance. Moreover, users might falsely assume the causes of irregular congestion and report a crash while simply stuck in traffic.

Creating a clean dataset by reconciling the variation in crowdsourced user reports of the same incident and matching these reports to incidents recorded in the ATMS represents one of the primary challenges. The matching procedures as explored in the literature are known as matching or conflation methods (*10–14*). As summarized Xavier et al. (*13*), similarity measures for point data (like the incident data in this study) are generally a combination of the following:

Geometric: Distance or area overlap.

Semantic: Measures of non-geometric properties.

Context: the special relationship between objects.

Ruiz et al. added the temporal criteria into their categories as well (*14*). For point matching, using geographic distance (Euclidian distance is most common) is the most classic approach (*15–17*). Adding extra information about the points when available, such as road names and direction, adds additional power to the matching function. The hybrid approach of geographic and semantic information has shown high accuracy in matching crowdsourced information (*18*). Considering the problem at hand and the available data in this research, a hybrid approach was used to leverage geographic as well as semantic matching methods.

Data

Waze Data

Waze is a navigation application that leverages crowdsourced user reports for providing service. Users can report traffic crashes, congestion, hazards, or police traps on the road (https://www.waze.com/about). The IDOT joined the Connected Citizen Program, which is an agreement in which the city or state managers provide Waze with information on road closures and constructions and, in return, Waze provides user reports to the managers. However, since the raw Waze data contain duplicate reports for a single incident and all reports may not have high reliability, data preprocessing is necessary (*19*). IDOT's ATMS implements stringent acceptance criteria for Waze reports before considering them for validation (filtering criteria: type = crash or reliability ≥ 6 or report rating ≥ 4). The reports that meet the criteria are sent to ATMS operators to verify the incident. If the incident is verified, it will be recorded in the ATMS database. However, in this work, all Waze reports were considered for comparison regardless of the reliability score.

ATMS Data

The Iowa ATMS records all incidents, hazards, and congestion detected by various sensors and cameras or the reports by the highway helpers or police. The incidents in this dataset are validated by ATMS operators and thus serve as a reference for evaluating other sources of data. However, not all incidents, particularly congestion, are recorded in this dataset.

Incidents Detected from Third-Party Traffic Services Vendors

Third-party traffic services vendors such as INRIX (www.inrix.com) gather anonymized position data, which in turn provide rural and urban system-wide traffic data with reasonable accuracy (*20, 21*). Iowa ATMS applies a state-of-the-art method for detecting incidents from INRIX data. This method utilizes interquartile range (IQR) of the historical speed data in each timeframe to detect outliers as described by Chakraborty and colleagues (*22*). Threshold speeds are computed for each segment, day of the week, and 15-minute period of the day, utilizing the last 8 weeks of data. More specifically, threshold = (Median − 2 × IQR) is computed for each period and an incident alarm is triggered when the real-time speed is below the corresponding threshold. The data generated from this process are another feed of data to the ATMS and a basis for comparison with Waze data.

Traffic Camera Images

Cameras mounted in various locations across Iowa are one of the main means for traffic monitoring in the ATMS. To estimate false alarms in Waze reports, this study uses screenshots of the camera video feed that are captured every 5 min. Cameras in the Des Moines, Iowa metropolitan area (56 cameras) were selected for manual labeling of road conditions. Since labeling the road conditions (particularly congestion) based on a single image is a subjective decision, the images were labeled "clear" when the road was obviously clear and no congestion or incidents were observed. The labeled road images were used to detect the false alarms in Waze reports.

Anticipated Coverage of Data Sources

In practice, each of these sources cover a portion of the true incidents; they have some overlaps, and may have false alarms as well. The Venn diagram of our data sources depicted in Figure 1 illustrates this relationship (circles are not drawn to scale); the characteristics of the overlapping areas are of primary interest. Iowa ATMS captures a subset of the true incidents which is validated and free from false alarms. Waze and INRIX are expected to cover some of the true incidents while having a portion of false alarms. This study is mainly focused on estimating the potential additional contribution of Waze to the ATMS (Region D). It is worth noting that the exact findings of this work are applicable to states and locations like Iowa, and that depending on the number of Waze users and penetration rates, the results may vary.

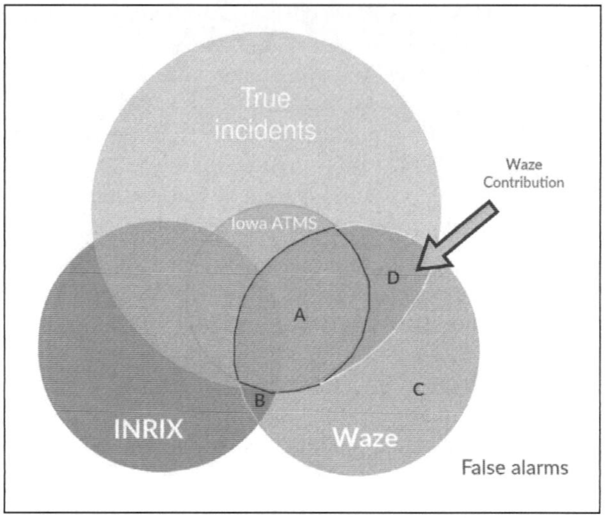

Figure 1. Venn diagram of the sources of traffic monitoring data, pointing to region of interest (D), the potential contribution of Waze.

Evaluation Procedure

Region (D) on the Venn diagram of our data sources (Figure 1) marks the potential contribution of Waze to the ATMS. However, since data on true incidents in all locations and times are not available, the existing sources were used to quantify the potential contribution and value in Waze feed. Therefore, the estimation of (D) was achieved in four main steps which are explained in this section using notations from Figure 1. The four steps are:

1. Match Waze and ATMS incidents (A)
2. Match Waze and INRIX incidents (B)
3. Estimate the false alarms (C)
4. Estimate Waze's contribution $D = \text{Waze} - (A \cup B \cup C)$

This study focused on two main types of incidents, congestion and crashes, as the sources that most directly impact traffic. To accomplish these steps, a matching function was necessary, which is described below.

Matching Function

For matching incidents between sources, a hybrid method leveraging geographic and semantic matching

Table 1. Event Matching Procedure for Step 1 (ATMS and Waze Matching)

Matching levels	Criterion	Logic	Matching method	Action category
First	Time	Waze reports 20 minutes before the start and after the end time of an ATMS record	Temporal	Preprocessing
Second	Location	Crashes in a 2.5-mile radius, Congestion in 1-mile radius	Geographic	Preprocessing
Third	Road name and direction	Grouped into: Matching both and opposite direction	Semantic	Preprocessing
Fourth	Type of incident	Type, road name, and direction match	Semantic	Full/exact Match
	Type of incident	ATMS event is a crash, Jam reported in Waze, No full match exists	Semantic	Secondary Jam of a crash
	Road direction	Everything matches, Opposite direction, 1-mile radius	Semantic	Opposite direction

Table 2. Summary of the Waze Evaluation Procedure Steps

Step	Name	Venn diagram segment	Research motivation	Data	
				Time	Location
0	Exploratory analysis	-	Waze and ATMS reports based on: Time of day, Region, Road type, Etc.	2016 entire year	entire state of Iowa
1	match Waze and ATMS	A	Waze and ATMS overlap, Redundancies, Influential factors in Waze coverage	2016 entire year	entire state of Iowa
2	match Waze and INRIX	B	ATMS and INRIX overlap, Waze vs. INRIX contribution to ATMS	October 2016	entire state of Iowa
3	Estimate the false alarms in Waze	C	% of Waze reports when road is clear (False alarms)	October 2016	Des Moines Area
4	Estimate Waze's contribution	D	The information that Waze can add		

methods was implemented. In both data sources, the road name and direction, as well as the type of the incident (i.e., crash, congestion, or stalled vehicle) were recorded. Table 1 presents the levels of the matching function as well as the criteria and method used in each level. The matching function first selects incidents in the temporal vicinity, then the geographic distance is examined. From spatiotemporal neighboring incidents, semantic information such as road names, direction, and type of the incident was used to mark matching incidents. The matching function introduced for this step (Match Waze and ATMS incidents) is the most comprehensive one. In the next steps, when matching with INRIX data and detecting false alarms, the match function was slightly modified to fit the semantic features of the respective data fields. Table 2 provides a summary of the evaluation procedure in this work and the data used in each step.

Results

Exploratory Waze Data Analysis

To initiate the evaluation, an exploratory data analysis was performed to better understand the Waze and ATMS data. The exploratory analysis looked into the pure number of reports regardless of the matching percentages or potential duplicates, to provide a high-level understanding of the two sources of data.

Sources of Incident Detection in the ATMS. Waze has been used as a source of incident detection in the IDOT

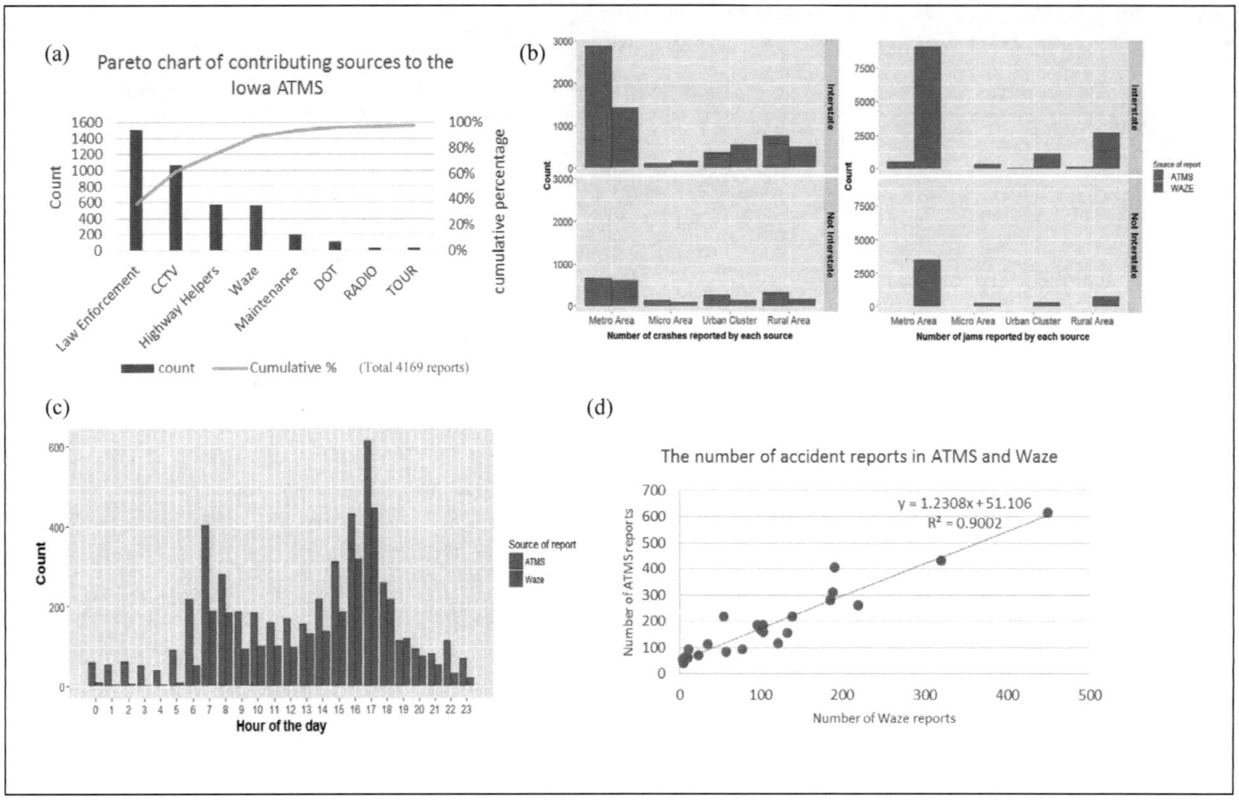

Figure 2. Exploratory data analysis results, comparing number of reports in Waze and ATMS. (*a*) A Parrateo chart of the portion of incidents detected from each source in Iowa ATMS records, indicating Waze as the fourth source of detection. (*b*) Visual comparison of the number of Waze and ATMS crash and congestion reports in multiple regions of the state as well as interstate and non-interstate roads. (*c*) The total number of crashes recorded in each source of data per hour of the day in year 2016 (only on weekdays). (*d*) The correlation between the number of crashes in each hour of the weekdays of year 2016 from each source (excluding midnight to 6 a.m.)

ATMS since September 2015. As depicted in Figure 2*a*, among the 23 sources of detection in the Iowa ATMS, law enforcement (which includes 911 calls, County Sheriff, State Patrol, etc.) contributes the highest number of incidents in the ATMS. Interestingly, Waze reports (detection source for 13.4% of the ATMS records) rank fourth in detection sources, after law enforcement, CCTV, and highway helpers. Comparing the operation and maintenance cost of each of the first three sources, Waze has a considerable contribution as a "free" detection source. However, in the current ATMS settings, the Waze reports need to be verified, usually by one of the top three sources, before being trusted.

Incident Reports in Distinct Locations and Road Types. The location of each report was mapped to the demographics of the region based on 2010 census data (*23*). Every county is grouped by their population as either metropolitan (>50,000), micropolitan (10,000–50,000), urban cluster (2,500–50,000), or rural (any non-urban region is considered rural). This analysis provides an insight into the spread and coverage of each source of data. As depicted in Figure 2*b*, the ATMS has recorded almost no congestion incidents (jams) outside of the metro area. although there are many congestion incidents reported in Waze from the urban clusters and rural areas (even off the interstates). In addition, the considerably larger numbers of reports on the interstates show the concentration of reports in both sources. This chart indicates the type of incident and locations where Waze could best contribute to the ATMS.

Impact of Time on the Waze Reports. To evaluate how the crowdsourced data reflect the reality on the roads, the number of reports in each hour of the day were compared and it was expected that the crowdsourced data would resemble the ATMS records. As observed in Figure 2*c*, both data sources tend to have a higher frequency of crash records during the rush hours. However, between midnight and 6 a.m., although ATMS shows 50–100 crash records, there are fewer than 10 Waze crash reports in the same time. The proportion of the number

Table 3. ATMS–Waze Matching Percentage by Report Type

Type of incident	Total reports in ATMS	% matched with Waze
Crashes	3,713	42.1
Congestion	456	58.5
Stalled vehicles	12,552	43.0

of Waze to ATMS crash reports during these hours (mean 9%) showed a statistically significant difference from the same proportion for other hours of the day (mean 37%). This indicates that Waze may not be a reliable detection source during midnight to 6 a.m. This observation aligns with the fact that during these hours there are fewer drivers on the roads and consequently fewer Waze users that might observe and report an incident.

Otherwise, the number of crashes reported in each hour of the day (from 6 a.m. to 11 p.m.) was highly correlated ($R^2 = 0.9$) between ATMS and Waze; as depicted in Figure 2*d*. Thus, the number of Waze crash reports during the day follows the reality of the roads.

Evaluation and Comparison

Step 1: The ATMS Incidents That Were Reported in Waze (Estimating A: Waze ∩ ATMS). This step compares Waze reports with the ATMS reports as source of validated events. The percentage of matching incidents in both sources answers questions regarding the reliability of Waze reports, while leading to the estimation of the potential contribution of Waze.

Using the described matching function, overall the congestion and crashes reported in Waze covered 43.2% of the ATMS records. The matching percentage by each type of incident is presented in Table 3.

In Iowa, similar to many other Midwestern U.S. states, traffic is not a daily concern for most people, and thus fewer people are familiar and active users of Waze, compared with more populated cities and states. Yet, the number of matched reports is interesting, considering a single crowdsourced feed of data has captured 43.2% of ATMS records.

What Factors Contribute to an Incident Being Reported in Waze in the Metro Area? To find the variables which have a statistically significant influence in determining whether an ATMS incident is reported in Waze, a binomial logistic regression was conducted. The binomial logistic regression was performed to ascertain the effects of day of the week, hour of the day, incident type, and the road type on the likelihood that an event covered by an ATMS record would be covered by Waze as well. The logistic regression model was statistically significant, $\chi^2(31) = 450.2$, $p << .001$. The model explained 20.0% (Nagelkerke R^2) of the variance in the matched instances and correctly classified 63.6% of cases. Of the 31 predictor variables (factors converted to dummy variables), the statistically significant ones were related to time and road type (as shown in Table 4). The incident type did not indicate a significant impact in this model.

As the road type turned out to be a significant contributing variable to the model, another logit model was tested using the interstate road names (nine variables) in the metro area as new variables, to investigate if a certain road significantly impacts the chance of an ATMS report being covered in Waze. None of the major interstates indicated a significant impact.

What Percentage of Waze was Covered in ATMS? And Were There Redundant Reports? Only 14.6% of the total Waze reports were matched with incidents in the ATMS

Table 4. Significant Influencers in ATMS–Waze Matching

Variable group	Variable	Estimate	*p*-value	Variable definition	
Time of the day	07:00–08:00	1.5444	< .0001**	07:00–09:00	Morning rush hour
	08:00–09:00	0.9143	.0003		
	11:00–12:00	0.6435	.0267	11:00–13:00	Lunch time
	12:00–13:00	0.7137	.0135		
	14:00–15:00	0.9792	.0004	14:00–19:00	Afternoon
	15:00–16:00	0.8815	.0006		
	16:00–17:00	1.5484	< .0001**		
	17:00–18:00	1.5602	< .0001**		
	18:00–19:00	0.8350	.0015		
	20:00–21:00	0.7376	.0333	20:00–21:00	Evening
Road type	Interstate or not	0.9083	< .0001**	Interstate/freeway or not**	

**indicates significance level of 0.001.

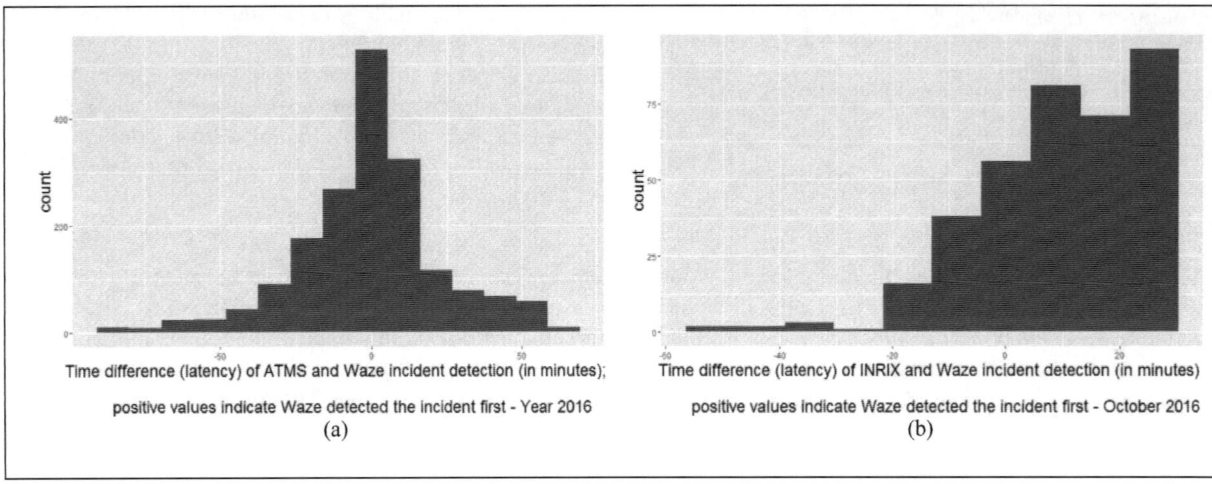

Figure 3. Waze incident detection time compared with ATMS and INRIX. (*a*) Time difference (latency) of ATMS and Waze incident detection (in minutes); positive values indicate Waze detected the incident first (Year 2016); (*b*) Time difference (latency) of INRIX and Waze incident detection (in minutes); positive values indicate Waze detected the incident first (October 2016).

records (36.8% for the crashes and 10.0% of the congestion). Thus, it is critical to investigate the unmatched Waze data to estimate the potential added coverage of Waze.

It was also found that on average, each ATMS report matched to 1.9 Waze reports, indicating the redundancy rate in Waze data. The median is 1 report, mean is 1.9, and 80% of the reports have two or fewer matches in Waze.

To examine the accuracy of the matching in distance, the 95% confidence interval for the distance between the matched Waze report and the ATMS record was calculated as .36–.39 miles. Evaluating the time accuracy of the matches, the time difference (latency of the reports) was calculated. As depicted in Figure 3*a*, the time difference forms a bell-shaped distribution around –0.22 minutes (95% CI, –1.3 to .8 minutes), which is slightly skewed to left. Slightly more than half of the matched incidents were detected earlier in Waze than the ATMS record.

Step 2: Estimating the Common Incidents in INRIX and Waze (B). Although the INRIX reports are not all validated, the overlap of Waze and INRIX reports increases the plausibility of an actual incident occurrence in the same time and location. To control for weather effects in our results, one month with relatively stable weather and about average matching percentages from Waze and ATMS incidents, was desired. October fulfilled the desired properties; therefore, October 2016 data was used for this part. Having applied incident detection method in Iowa ATMS, as described by Chakraborty et al. (*22*), the incidents were detected from INRIX.

Using the described matching function in Table 1, 48% of Region A of Figure 1 (Waze ∩ ATMS) was also matched with INRIX. This result implies that the INRIX feed had detected about half of the common incidents in Waze and ATMS, adding to the validity of the INRIX-detected incidents.

To estimate Region B on the Venn diagram, the overlap of the Waze reports with the INRIX data was evaluated. The results indicated 16.8% of Waze reports were matched to INRIX. The time difference between Waze reports and matched incidents demonstrated that on average, INRIX reports were detected 9.8 minutes later (95% CI, 8.25–11.36) than Waze reports (Figure 3*b*).

Step 3: Estimating the False Alarms in the Metro Area (C). Region C of Figure 1 represents false alarms from Waze, that is, reports of incidents that did not actually exist. To estimate the number of false alarms in Waze, manually labeled images from IDOT cameras in the Des Moines metro area were used. The results indicated that overall, only one of the 319 Waze reports in October 2016 and locations was a false alarm. This accounts for 0.3% of the reports.

Although our false alarm definition is not strict (a false alarm is when the road is visibly clear and there is a Waze incident report), the false alarm rate is interestingly lower than expectations. It is worth mentioning a great portion of Waze reports are congestion reports that DOT is not particularly interested in recording. Yet, this is an important finding to understand the validity of these crowdsourced Waze reports.

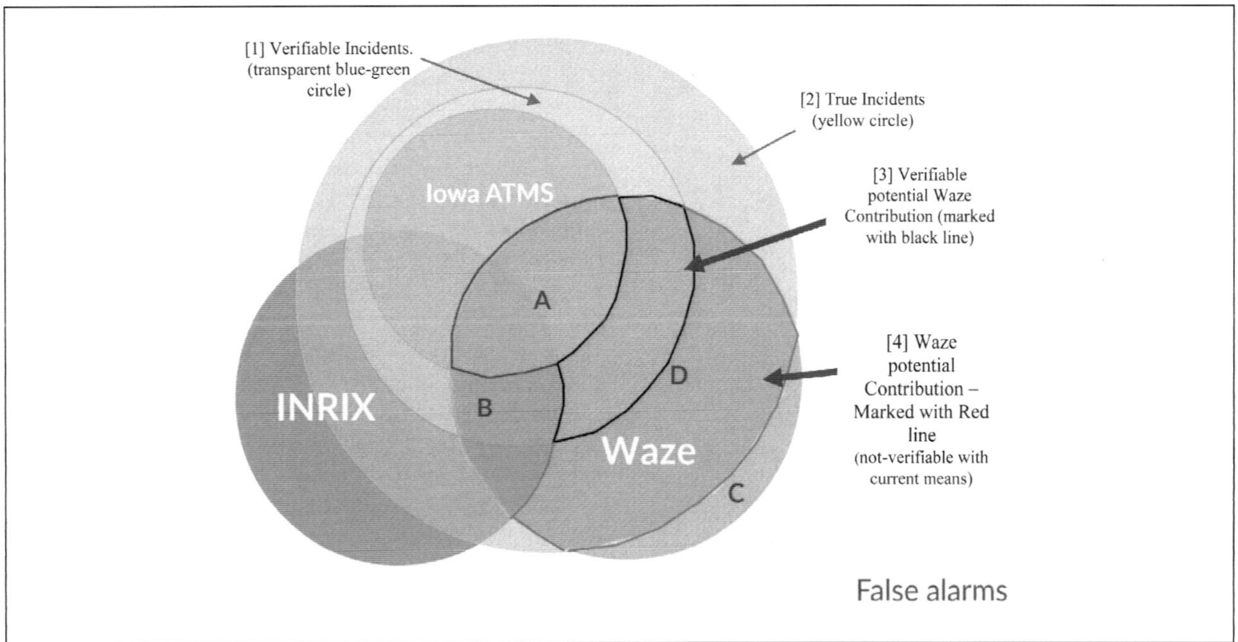

Figure 4. Updated Venn diagram based on the analysis, the regions are drawn closer to scale. Region D, the estimated contribution of Waze to the ATMS, is divided into verifiable and non-verifiable regions.

Step 4: Estimating the Waze Contribution (D). The final step in the process is to estimate the Waze contribution, or Region D on the Venn diagram of Figure 1. Based on the following calculations, 68.3% of the Waze incidents were estimated to be the additional information that Waze can contribute. Once accounting for the number of redundant reports (1.9 redundant reports was rounded up to 2.0 for a more conservative estimation), 34.1% of the Waze's crash and congestion reports (7387 instances which are mainly congestion reports) were potential incidents that were not recorded by the current sources of the ATMS.

$$D = (A \cup B \cup C)'$$
$$= 100\% - (14.6\% + 16.8\% + 0.3\%) = 68.3\%$$

$$Accounting\ for\ redundancies : \frac{68.3\%}{2} = 34.1\%$$

$$Number\ of\ incidents : 34.1\%$$
$$(total\ congestion\ and\ crashreports\ in\ Waze) =$$
$$.341 \times 21662 \cong 7387$$

To further estimate the potential additional crash coverage in Waze data, the proportion of crash reports among Region D within the 2016 data was 12% of all Region D incidents. Assuming this percentage is uniform in the unmatched Waze reports, this yields about 904 crashes in year 2016 (12% × 7387 reports) which are either potentially missed or recorded with different labels by the ATMS. These numbers provide an estimate of Waze's potential contribution to traffic coverage in the state of Iowa.

Note that the Waze congestion reports do not come with the recurring or non-recurring labels. Thus, many of the congestion reports might be recurring traffic patterns. Although the ATMS operators are not concerned with recurrent congestion, the Waze reports still provide invaluable information about the traffic conditions. Moreover, records on all types of traffic incidents provide training data for classification models that can distinguish recurring and non-recurring congestion.

Comparing Waze with Findings about Twitter. Now that the contribution of Waze has been estimated, it is worth examining its performance with other data sources of data. The work of Gu et al. (*2*) provided information about traffic incidents extracted from Twitter in Pennsylvania. Comparing some of the findings about Twitter with Waze was insightful. Like the present results with Waze, Gu et al.'s analysis showed Twitter to be less reliable during night hours. Also, most of the tweets were during the peak traffic hours. Gu et al. reported an average of 1.6 Twitter-reported incidents per unique incident. This number was estimated 1.9 reports for Waze, indicating that redundant reports are a common challenge in other crowdsourced data feeds.

Summary of the Findings

Based on the quantitative analysis of Waze data, Figure 4 is an updated view of the Venn diagram that better illustrates the relationship and overlap of the three sources of data. In this another aspect of the challenge is demonstrated. Although there exists a set of true incidents (the yellow circle), not all of them are known through the existing means. Thus, when evaluating the potential of Waze this challenge should be acknowledged. Note that the (D) region in the figure is now split into sections [3] and [4]. The overlap of (D) and Verifiable incidents [3] shows the incidents that are verifiable through other existing means (particularly CCTV cameras). Part [4] in region (D) are reports that can potentially be valid incidents, and there are currently no cameras or other means to verify their accuracy. Based on this work, it is believed that a considerable percentage of the potential incidents in (D) provide invaluable information to the ATMS.

Discussion and Conclusion

This research evaluated crowdsourced traffic incident reports from Waze, to study its characteristics as a data source. This section provides a summary of the findings.

How Does Waze Compare with the Existing Sources?

The reliability of crowdsourced incident reports from Waze was affirmed with the matching percentages between Waze and validated ATMS (42.3% of ATMS records) and INRIX data. In the Iowa ATMS, 13.4% of the recorded congestion and crashes were initially detected by Waze reports, making it the fourth most contributing source of incident detection. These findings indicate the reliability and competent coverage of crowdsourced traffic incident reports such as Waze.

What are the Characteristics of Waze Data?

Waze incident reports indicated a widespread coverage of instances in most locations and road types, particularly for reported congestion. The quality of the reports did not depend on the day of week or a specific roadway. On the other hand, the analysis indicated in the less crowded hours of the day (12 a.m. to 6 a.m.), Waze reports are not a reliable source for monitoring road conditions.

What is the Estimated Potential Additional Coverage that Waze can Provide to the ATMS?

The potential additional coverage that Waze can provide to the ATMS was estimated to be 34.1% of Waze reports, which accounts for 7387 incidents per year (from which 904 were estimated to be crash reports), making it a valuable source for traffic managers to utilize.

Overall, it can be concluded that crowdsourced reports such as Waze are invaluable sources of information for traffic monitoring with broad coverage, timely response time, and reasonable accuracy. Integrating this source of data into the ATMS feeds provides significant contributions to the traffic monitoring coverage.

However, there are challenges in working with this crowd-based data, including redundancies, inaccuracies, and mismatches in report types, as well as the need for report reliability estimation. Therefore, preprocessing and validating such data is necessary and requires resource investment. The crowdsourced data, on the other hand, are typically provided freely (or at a low cost) to the ATMS managers. Compared with the immense cost of installation and maintenance of other data sources (sensors, third-party probe data, or even law enforcement reports), raw Waze data are available for free. This analysis indicated potential valuable incident information from cleaned and processed Waze data. Therefore, a short-term investment in human resources to establish an infrastructure for eliciting valuable information from Waze data seems economically justifiable. This infrastructure would include models to address the redundancy issue and to automatically estimate the reliability of the reports, which are directions for future work.

Although the exact value of Waze data would vary for different regions and over time, these numbers in a less congested U.S. state seem impressive, and the techniques used in this research for Waze data evaluation could be applied to any region. Moreover, knowing the number of active Waze users in different regions would add a valuable basis for comparative across multiple regions.

Acknowledgments

This material is based on the work funded by Iowa Department of Transportation.

Author Contributions

The authors confirm contribution to the paper as follows: study conception and design: Amin-Naseri, Sharma, Gilbert, and Hong; data collection: Amin-Naseri and Chakraborty; analysis and interpretation of results: Amin-Naseri, Sharma, Gilbert, and Hong; draft manuscript preparation: Amin-Naseri, Gilbert, and Chakraborty. All authors reviewed the results and approved the final version of the manuscript.

References

1. Yoon, J., B. Noble, and M. Liu. Surface Street Traffic Estimation. *Proc., 5th International Conference on Mobile Systems, Applications, and Services, ACM*, New York, 2007, pp. 220–232.

2. Gu, Y., Z. (Sean) Qian, and F. Chen. From Twitter to Detector: Real-Time Traffic Incident Detection Using Social Media Data. *Transportation Research Part C: Emerging Technologies*, Vol. 67, 2016, pp. 321–342.
3. Li, R., K. H. Lei, R. Khadiwala, and K. C. C. Chang. TEDAS: A Twitter-Based Event Detection and Analysis System. *Proc., 28th International Conference on Data Engineering*, IEEE, Washington, D.C., IEEE, 2012, pp. 1273–1276.
4. D'Andrea, E., P. Ducange, B. Lazzerini, and F. Marcelloni. Real-Time Detection of Traffic from Twitter Stream Analysis. *IEEE Transactions on Intelligent Transportation Systems*, Vol. 16, No. 4, 2015, pp. 2269–2283.
5. Seeger, C., C. Lillehoj, S. Wilson, and A. Jensen. Facilitated-VGI, Smartphones and Geodesign: Building a Coalition while Mapping Community Infrastructure. Proc., *Digital Landscape Architecture 2014 - Landscape Architecture and Planning: Developing Digital Methods in GeoDesign*, Herbert Wichmann Verlag, Zurich, 2014, pp. 300–308.
6. Sakaki, T., M. Okazaki, and Y. Matsuo. Earthquake Shakes Twitter Users: Real-time Event Detection by Social Sensors. *Proc., 19th International Conference on World Wide Web*, ACM, New York, 2010, pp. 851–860.
7. Aramaki, E. Twitter Catches The Flu: Detecting Influenza Epidemics using Twitter. *Computational Linguistics*, Vol. 2011, 2011, pp. 1568–1576.
8. Signorini, A., A. M. Segre, and P. M. Polgreen. The Use of Twitter to Track Levels of Disease Activity and Public Concern in The U.S. during the Influenza A H1N1 Pandemic. *PLoS One*, Vol. 6, No. 5, 2011, e19467.
9. Steiger, E., B. Resch, J. P. de Albuquerque, and A. Zipf. Mining and Correlating Traffic Events from Human Sensor Observations with Official Transport Data Using Self-Organizing-Maps. *Transportation Research Part C: Emerging Technologies*, Vol. 73, 2016, pp. 91–104.
10. Goh, C. Y., J. Dauwels, N. Mitrovic, M. T. Asif, A. Oran, and P. Jaillet. Online Map-Matching Based on Hidden Markov Model for Real-Time Traffic Sensing Applications. *Proc., 15th International IEEE Conference on Intelligent Transportation Systems (ITSC), IEEE*, Anchorage, Alaska, Vol. 117543, 2012, pp. 776–781.
11. Sester, M., J. J. Arsanjani, R. Klammer, D. Burghardt, and J.-H. Haunert. Integrating and Generalising Volunteered Geographic Information. In *Abstracting Geographic Information in a Data Rich World* (D. Burghardt, C. Duchene, and W. Mackaness eds.), Springer International Publishing, Cham, 2014, pp. 119–155.
12. Yang, B., and Y. Zhang. Pattern-Mining Approach for Conflating Crowdsourcing Road Networks with POIs. *International Journal of Geographical Information Science*, Vol. 8816, No. June, 2015, pp. 1–20.
13. Xavier, E. M. A., F. J. Ariza-López, and M. A. Ureña-Cámara. A Survey of Measures and Methods for Matching Geospatial Vector Datasets. *ACM Computing Surveys*, Vol. 49, No. 2, 2016, p. Article 39.
14. Ruiz, J. J., F. J. Ariza, M. a. Ureña, and E. B. Blázquez. Digital Map Conflation: A Review of The Process and a Proposal for Classification. *International Journal of Geographical Information Science*, Vol. 25, No. 9, 2011, pp. 1439–1466.
15. Beeri, C., Y. Kanza, E. Safra, and Y. Sagiv. Object Fusion in Geographic Information Systems. *Proc., 13th International Conference on Very Large Data Bases*, Vol. 30, 2004, pp. 816–827.
16. Safra, E., Y. Kanza, Y. Sagiv, C. Beeri, and Y. Doytsher. Location-Based Algorithms for Finding Sets of Corresponding Objects over Several Geo-Spatial Data Sets. *International Journal of Geographical Information Science*, Vol. 24, No. 1, 2010, pp. 69–106.
17. Beeri, C., Y. Doytsher, Y. Kanza, E. Safra, and Y. Sagiv. Finding Corresponding Objects when Integrating Several Geo-Spatial Datasets. *Proc., 13th Annual ACM International Workshop on Geographic Information Systems*, ACM, New York, 2005, pp. 87–96.
18. McKenzie, G., K. Janowicz, and B. Adams. A Weighted Multi-Attribute Method for Matching User-Generated Points of Interest. *Cartography and Geographic Information Science*, Vol. 41, No. 2, 2014, pp. 125–137.
19. Pack, B. M., and N. Ivanov. Are You Going My Waze? Practical Advice for Working. *ITEjournal*, Vol. 87, No. 2, 2017, pp. 28–35.
20. Li, H., S. M. Remias, C. M. Day, M. M. Mekker, J. R. Sturdevant, and D. M. Bullock. Shock Wave Boundary Identification Using Cloud-Based Probe Data. *Transportation Research Record: Journal of the Transportation Research Board*, No. 2526, 2015, pp. 51–60.
21. Haghani, A., M. Hamedi, and K. F. Sadabadi. Summary report of I-95 Corridor Coalition Vehicle Probe Project: Validation of INRIX Data. *I-95 Corridor Coalition*, 2009, p. 7.
22. Chakraborty, P., J. R. Hess, A. Sharma, and S. Knickerbocker. Outlier Mining Based Traffic Incident Detection Using Big Data Analytics. Presented at 96th Annual Meeting of the Transportation Research Board, Washington, D.C., 2017.
23. United States Census Bureau. *Urban Area Relationship Files*. https://www.census.gov/geo/maps-data/data/ua_rel_download.html Accessed January 01, 2017.

The Standing Committee on Highway Traffic Monitoring (ABJ35) peer-reviewed this paper (18-03519).

Any opinions, findings, and conclusions or recommendations expressed in this material are those of the authors and do not necessarily reflect the views of Iowa Department of transportation.

Article

TRR
JOURNAL OF THE TRANSPORTATION RESEARCH BOARD

Transportation Research Record
2018, Vol. 2672(43) 44–54

DOI: 10.1177/0361198118774735
journals.sagepub.com/home/trr
SAGE

Method for Imputing Missing Data using Online Calibration for Urban Freeway Control

Xu Wang[1], Yuechun Ge[1], Lei Niu[1], Yi He[2], and Tony Z. Qiu[3]

Abstract

Real-time traffic control systems are widely implemented on roadways around the world as a measure to improve freeway mobility. However, the systems, which rely on data from road-side and on-road sensors and other electronic equipment, continue to suffer from issues related to missing and erroneous data. While many data imputation methods are documented in the related literature, traffic control systems still lack an imputation method that is applicable in practice, accurate in imputation, and simple in computation. In response, this paper puts forth a linear imputation model that considers both temporal traffic trend and spatial detector correlations. To adapt the model to dynamic traffic variations, the imputation method was equipped with an online calibration module. The proposed imputation method was evaluated with field data from two stations on the Whitemud Drive, a busy urban freeway in Edmonton, Alberta, Canada. The proposed model benefited from its time-of-day temporal trend and outperforms the previous model that considers only spatial correlations. Moreover, the online calibration module was effective in improving imputation accuracy. Finally, the sensitivity of imputation performance was analyzed. The results show that the imputation with online calibration is more sensitive to missing data ratios than that with offline calibration. The sensitivity analysis revealed that imputation with online calibration is more suitable for online imputation in traffic control implementations.

Worldwide, urban freeways experience congestion caused by ever-increasing traffic demands. In attempts to combat such congestion, intelligent transportation systems for freeway control have been designed and implemented in real-world applications (*1–3*). These traffic control systems collect data from sensors, estimate or predict traffic evolutions, and optimize and implement control decisions. Freeway control systems rely heavily on real-time traffic data. However, data collected in real time are highly prone to experiencing issues related to missing or erroneous data as a result of hardware or communication failure. Traffic control systems require accurate and complete data to operate efficiently and improve freeway mobility. If traffic data are missing for even a short time period, the traffic control terminates, temporarily leading to unreliable system performance and potentially threatening traffic safety. Thus, before applying control strategies in the field, imputation methods that address missing and erroneous data issues must be designed to improve data quality. This study aimed to do just that.

Traffic detectors on urban freeways collect a vast amount of data every day. Real-time freeway control requires practical, accurate, and time-efficient imputation methods, because storage space and processing time are limited. Existing studies have developed imputation methods for online applications. These online imputation methods model either only temporal or only spatial relationships by: (1) replacing missing data with historical time-of-day mean values, which reflects the daily traffic trend; or (2) imputing missing data using the spatial relationships with surrounding detectors, which supplements real-time fluctuations. However, few online imputation methods combine both temporal trend and spatial relationships. Imputation with only one relationship cannot capture the daily trend and real-time variations simultaneously. Even worse, imputation that only applies spatial

[1]School of Civil Engineering, Shandong University, China
[2]Engineering Research Center for Transportation Safety (Ministry of Education), Wuhan University of Technology, Wuhan, China
[3]Department of Civil and Environmental Engineering, University of Alberta, Edmonton, AB, Canada

Corresponding Author:
Address correspondence to Yi He: heyi@whut.edu.cn

relationships may fail when missing problems occasionally happen on all the surrounding detectors. Moreover, previous online imputation methods apply static model parameters, which may misestimate the temporal and spatial traffic variations. To bridge the research gaps, this study created an imputation method with online calibration that considers both temporal and spatial relationships. This approach is promising in real-world applications because of its accuracy, computational simplicity, and small storage space requirements.

The remainder of this paper is organized into sections: the next section reviews the existing studies; the Methodology section details the proposed method of this study; the Imputation Results section deploys and evaluates the assessed imputation method performance; and finally, the last section discusses the concluding remarks and suggests future work.

Literature Review

Traffic data imputation methods have long been investigated. Most methods use historical data or the neighboring functioning detector data to estimate any missing values. There are two major purposes for data imputation in previous research: (1) data archival for further analysis and (2) real-time applications (*4*). Smith et al. compared heuristic and statistical imputation methods and indicated that data imputation methods differ in computational complexity, speed, and accuracy (*4*). Thus, imputation for data archival usually uses sophisticated offline methods, while methods for real-time applications use heuristic online techniques. The offline methods utilize the information before and after the missing data points. Inductive learning methods that capture stochastic traffic characteristics succeed in offline methods. Such offline methods include Markov chain Monte Carlo (*5*), probabilistic principal component analysis (*6*), matrix completion (*7*), fuzzy C-means (*8*), and tensor-based (*9*) methods. Other methods search for close-in-pattern historical data, for example *k*-nearest neighbors (*10*) and time-series (*11*) methods.

In contrast to offline methods, online imputation methods occupy a smaller storage space and take less processing time. In practical implementation, the data size is so large that it is impractical to store significant historical data. Also, online methods generate suitable estimations and require only one data collection interval to replace missing data. Most imputation methods applicable for online applications are regression based. For example, Chen et al. estimated missing data based on the spatial pairwise relationship between neighboring loops (*12*). Therefore, the proposed algorithm here made full use of available data from neighboring sensors. Henrickson et al. developed a proven predictive mean-matching multiple imputation method (*13*). They quantified imputation accuracy in various missing data scenarios. Bie et al. found a linear relationship among detectors at the same station and constructed a multiple linear regression model to fill in the missing data (*14*). Yin et al. examined the performance of temporal and spatial imputation methods (*15*). They found that the method that substitutes the time-of-day historical mean for a missing value outperforms others in accuracy, but it requires more data archiving and a longer processing time. They also pointed out that the performance of spatial methods is sensitive to lane configurations.

The model parameters in the online imputation methods above were all predetermined and remained static during implementation. In contrast, Smith and Conklin (*16*) observed that there exists a consistent lane distribution pattern by location and time of day. Based on this consistency, they developed a data estimation procedure and obtained acceptable estimation accuracy. It was found that the relationship between traffic measurements among lanes is time variant and lane specific. Therefore, imputation methods that apply static model parameters may not reflect the real-time temporal and spatial relationships. In addition, the parameters to be adjusted in the methods should be kept to as few as possible, and should be made easy to understand for operational staff.

Previous studies also do not determine the maximum missing data ratio that imputation methods can withstand. Without this determination, the imputation errors accumulate and control systems must be terminated. Considering the requirements for online imputation methods and the limitation of existing research, development of imputation methods with online calibration is an option to bridge this research gap.

Online calibration adjusts model parameters based on real-time data so that the parameters can reflect current states. Many online calibration methods have been applied to calibrate macroscopic (*17*) and microscopic (*18*) models. The commonly used approaches for online calibration are the extensions of Kalman Filter. Existing research proves that online calibration generates more reliable traffic evolutions, especially for cases with poor historical information or traffic behavior variation (*17*, *18*). The approach proposed by Papathanasopoulou et al. included an estimation phase and a prediction phase (*18*). It was demonstrated as a practical approach for online calibration of a microscopic traffic prediction model.

The present study involved an online calibration approach along with an online imputation approach to create a practical imputation method for freeway control systems. There were three objectives of this study: (1) develop an imputation model that considers temporal trend and spatial correlations as well as online calibration that is potentially applicable for practice; (2)

evaluate the proposed method with real-world traffic data; and (3) investigate the sensitivity of imputation accuracy by missing data ratios. This paper presents the resulting multiple linear regression model with online calibration. The evaluation was conducted with field traffic data gathered from Whitemud Drive in Edmonton, Alberta, Canada. The proposed imputation method may lead to the implementation of urban freeway control systems.

Methodology

Imputation Procedure

This study deployed a three-step procedure to impute missing data in real-time (Figure 1). To apply the imputation procedure, the data collection stations were indexed as $m = 1, \ldots, M$. At one station, the detectors for each lane were indexed as $n = 1, \ldots, N$. When the imputation procedure activated, it diagnosed data collection problems at each time index t ($t = 1, \ldots, T$). If traffic data were detected as missing or erroneous, the imputation and online calibration algorithms activated. The online calibration estimated model parameters by optimizing traffic measurements at the preceding time steps. Next, it predicted parameters for the current time step and input the prediction results into the imputation module. With all of that information in hand, the imputation problem was established as a multi-linear regression problem. The imputation results updated the database for the next imputations.

Diagnostics Algorithm

This study detected missing and erroneous data by applying the diagnostic method developed in previous research (*14*). The diagnostic algorithm was based on the measurement thresholds (*19*), acceptable regions (*20*), and corresponding upstream data. To be specific, the algorithm pre-determines thresholds and feasible density-flow regions for traffic measurements. Once real-time collected traffic measurements fall beyond the thresholds or feasible regions, the algorithm reports invalid traffic measurements.

Traffic control works mostly during peak hours when there is a greater possibility of non-zero traffic measurements for a data collection interval larger than 1 minute. If a detector reports a zero flow but its corresponding upstream lane has a non-zero flow, the traffic measurements are diagnosed as missing data. It is important to note that the diagnostics algorithm ignored the most upstream station, because it had no upstream station. The diagnostics algorithm generated a data completeness matrix Δ determined by the following equation. The online calibration and imputation modules activated for the lane(s) where $\Delta = 0$. In the following discussion, the term "missing data" refers to missing and erroneous data, unless otherwise noted.

$$\Delta_{t,n}^{m} = \begin{cases} 0, \text{if data are diagnosed as missing or erroneous} \\ 1, \text{otherwise} \end{cases} \tag{1}$$

Imputation Model

This study proposed an imputation method based on a multi-linear regression model. Traffic data, especially flow and density, are correlated with the time-of-day historical mean (*15*) and its adjacent detectors (*12*, *15*, *16*). The proposed model considers both temporal trend and spatial relationships. The time-of-day historical mean values reflect the general temporal trend, while the

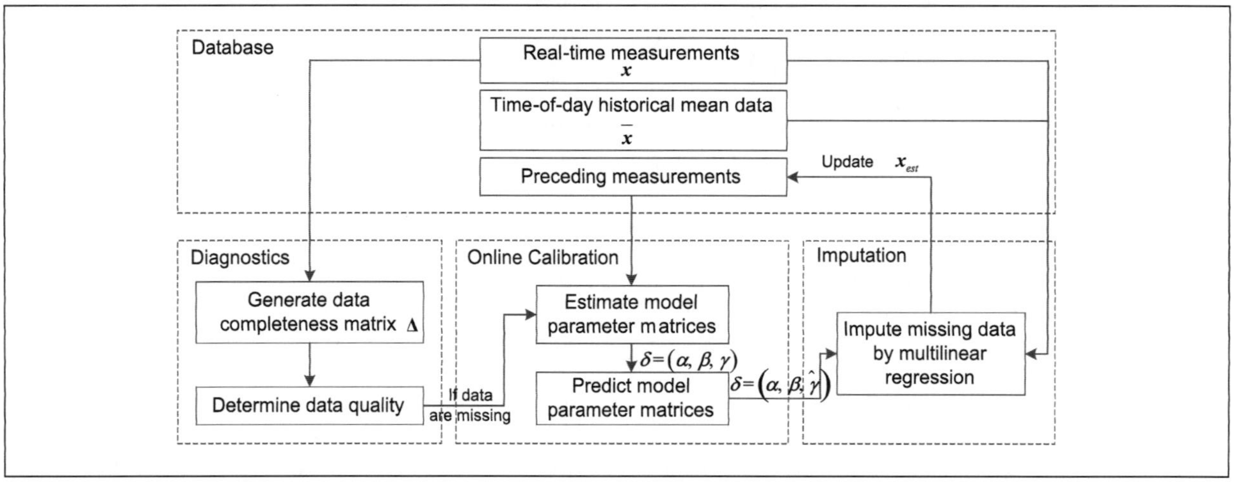

Figure 1. Methodology flowchart.

estimations from the relationship among detectors show the immediate fluctuations. The proposed imputation algorithm combines temporal trend with spatial immediate fluctuation to obtain better imputation results. The imputed traffic data x_{est} was formulated as follows:

$$x^{m}_{\text{est},t} = \Delta^{m}_{t} x^{m}_{t} + (\Omega - \Delta^{m}_{t}) \left(\widehat{\alpha}^{m}_{t} \Delta^{m}_{t} x^{m}_{t} + \widehat{\beta}^{m}_{t} \overline{x}^{m}_{t} + \widehat{\gamma}^{m}_{t} + \varepsilon^{m}_{t}\right) \tag{2}$$

where,

$$x = (x_1, \ldots, x_N)^T \tag{3}$$

$$\overline{x} = (\overline{x}_1, \ldots, \overline{x}_N)^T \tag{4}$$

$$\widehat{\alpha} = \begin{pmatrix} \widehat{\alpha}_{11} & \cdots & \widehat{\alpha}_{1N} \\ \vdots & \ddots & \vdots \\ \widehat{\alpha}_{N1} & \cdots & \widehat{\alpha}_{NN} \end{pmatrix} \tag{5}$$

$$\widehat{\beta} = diag\left(\widehat{\beta}_1, \ldots, \widehat{\beta}_N\right) \tag{6}$$

$$\widehat{\gamma} = (\widehat{\gamma}_0, \ldots, \widehat{\gamma}_N)^T \tag{7}$$

$$\Delta = diag(\Delta_1, \ldots, \Delta_N) \tag{8}$$

$$\Omega = diag(1, \ldots, 1) \tag{9}$$

$$\varepsilon = (\varepsilon_0, \ldots, \varepsilon_N)^T \tag{10}$$

In the imputation model, x is flow or density measurements from both malfunctioning and functioning detectors as shown in Equation (3); $\overline{x}$ is time-of-day historical mean of flow or density measurements at the imputed detector (see Equation [4]); $x_1, \ldots, x_N$ and $\overline{x}_1, \ldots, \overline{x}_N$ are traffic measurements and time-of-day historical mean values at lane 1 to lane N at one station. The combination of temporal trend and spatial relationships ensures that the proposed algorithm can estimate the omitted values in all missing data situations. The estimation from time-of-day historical mean can fill the missing values even though data from all the surrounding detectors are missing.

Parameters $\widehat{\alpha}$, $\widehat{\beta}$ and $\widehat{\gamma}$ are the model parameters predicted in the online calibration phase ($\widehat{\delta} = \left(\widehat{\alpha}, \widehat{\beta}, \widehat{\gamma}\right)$); $\widehat{\alpha}$ and $\widehat{\beta}$ are shown in Equations (5) and (6); $\alpha_{nn'}$ is the model parameter of the relationship between lane n and lane n' for data imputation on lane n; β_n is the model parameter of the time-of-day historical mean on lane n for data imputation on lane n; $\widehat{\gamma}$ denotes the constant term; Δ denotes the data completeness matrix as shown in Equation (8), which is generated from the diagnostics phase ($\Delta_1, \ldots, \Delta_N = \begin{cases} 0, \text{missingdata} \\ 1, \text{completedata} \end{cases}$); and ε is the error term.

Online Calibration

The online calibration approach had two phases. First, the estimation phase optimized the model parameters ($\delta = (\alpha, \beta, \gamma)$) in the following equation using data from the preceding time steps. The length of preceding time steps i was set at 5 minutes in this study. It is because from the observations in Smith and Conklin (*16*), the lane distribution is steady at a time interval less than 10 minutes. Second, the prediction phase forecasted model parameters for the current time step based on the estimation results from the previous time steps.

$$x^{m}_{\text{est},t} = \Delta^{m}_{t} x^{m}_{t} + (\Omega - \Delta^{m}_{t})(\alpha^{m}_{t} \Delta^{m}_{t} x^{m}_{t} + \beta^{m}_{t} \overline{x}^{m}_{t} + \gamma^{m}_{t} + \varepsilon^{m}_{t}) \tag{11}$$

The estimation problem was mathematically solved by the least square regression, so the error terms ε can be treated as zero. By repeating this optimization process, the estimation for the previous time steps was obtained. The prediction phase applied an autoregressive process to predict $\widehat{\delta}$. The way to determine the factors is explained by Antoniou et al. (*21*).

$$\widehat{\delta}_t = w_1 \delta_{t-1} + w_2 \delta_{t-2} + \ldots w_i \delta_{t-i} + \varepsilon \tag{12}$$

where w are the autoregressive factors and ε is a random error term.

Imputation Results

Study Site

The City of Edmonton installed vehicle detection stations (VDSs) and traffic video cameras on a busy urban freeway corridor called the Whitemud Drive (WMD) in Edmonton, Alberta, Canada. Data from two stations, VDS 1029 and VDS 1031, were analyzed to evaluate the proposed imputation method.

Figure 2 schematically shows the lane configurations and detector locations. VDS 1029 is followed by a left-hand off-ramp to Terwillegar Drive. The lane number on the mainline decreases from three to two, and most vehicles travel in the mainline instead of the off-ramp during peak hours. Thus, congestion often begins at this segment and propagates far upstream on weekday peak hours. VDS 1031 is located on a one-sided weaving segment, originating at the onramp of 53 Avenue to the off-ramp at Fox Drive. The VDSs collect traffic data on volume, speed, and occupancy every 20 seconds (s), and send the data to the city's central computer system for archive. The real-time data are aggregated into 1-minute intervals and input into traffic control applications, for example, variable speed limit control (*3*).

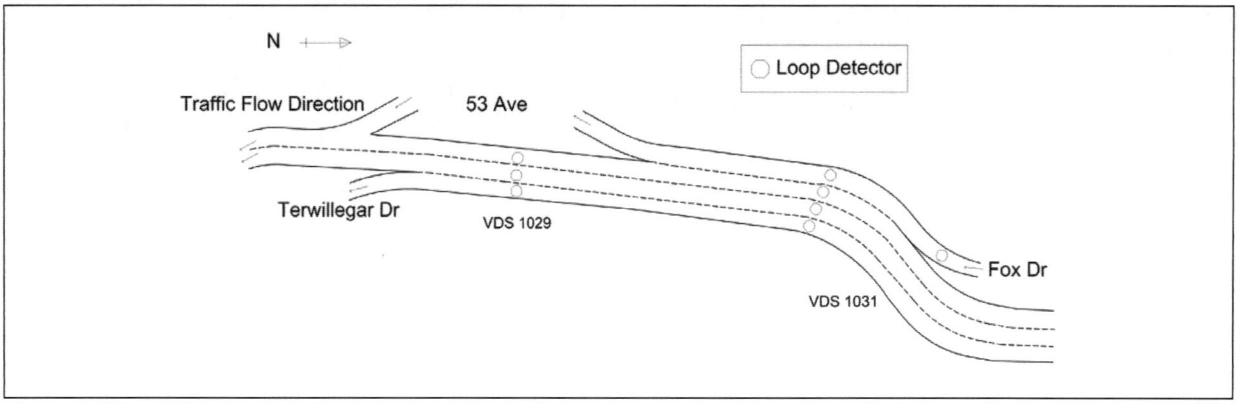

Figure 2. Study site.

Table 1. Pearson Correlation Coefficients between Lanes

	Lane no.	Lane no.	Flow	Density	Speed
VDS 1029	1	2	0.9728	0.9753	0.6380
	2	3	0.7664	0.7168	0.2817
	1	3	0.8042	0.7507	0.2523
VDS 1031	1	2	0.7669	0.6997	0.0431
	2	3	0.9654	0.9644	0.8527
	3	4	0.8882	0.8846	0.4723
	1	3	0.7678	0.7502	0.0083
	1	4	0.9049	0.8889	0.1089
	2	4	0.8838	0.8555	0.4610

Field Data Analysis

Complete datasets from VDSs were available from the years 2013 to 2016. Data under adverse weather and incident conditions were excluded to eliminate weather and incident impact. Table 1 lists the Pearson correlation coefficients of flow and density between lanes on VDS 1029 and VDS 1031 using the 4-year data. Flow and density were shown to be much more correlated than speed, so in this study, flow and density were imputed by the imputation model, but speed was then calculated from the relationship between flow, density, and speed. On VDS 1029, flow and density between lane 1 and lane 2 are more correlated than those between the other lanes. Similarly, on VDS 1031, correlations of flow and density between lane 2 and lane 3 and between lane 1 and lane 4 are much stronger than those between others. These correlations can also be observed in Figure 3. Figure 3 gives the flow and density correlation plots between lanes in May 2015. On VDS 1029, the correlation between lane 2 and lane 3 and between lane 1 and lane 3 is in an "italic V" shape. The observations of correlation can be readily explained: the flows on the lanes connected to the ramps can be remarkably different to those connected to the mainline.

Measures of Effectiveness

This study selected two measures of effectiveness (MOEs) to quantify the imputation performance: (1) mean absolute percent error (MAPE), which indicates how far the scaled imputes are from the actual values; and (2) normalized root mean square error (NRMSE), which evaluates the normalized overall estimation error and applies a much heavier weight to large errors than to small errors. These MOEs were calculated by the following equations:

$$MAPE = \frac{1}{T}\sum_{t=1}^{T}\left|\frac{x_{\mathrm{est,t}} - x_t}{x_t}\right| \tag{13}$$

$$NRMSE = \frac{\sqrt{T \cdot \sum_{t=1}^{T} [x_{\mathrm{est,t}} - x_t]^2}}{\sum_{t=1}^{T} x_t} \tag{14}$$

Model Results

The imputation method was coded and implemented in MATLAB software (*22*). The evaluation application created data completeness matrices and removed the

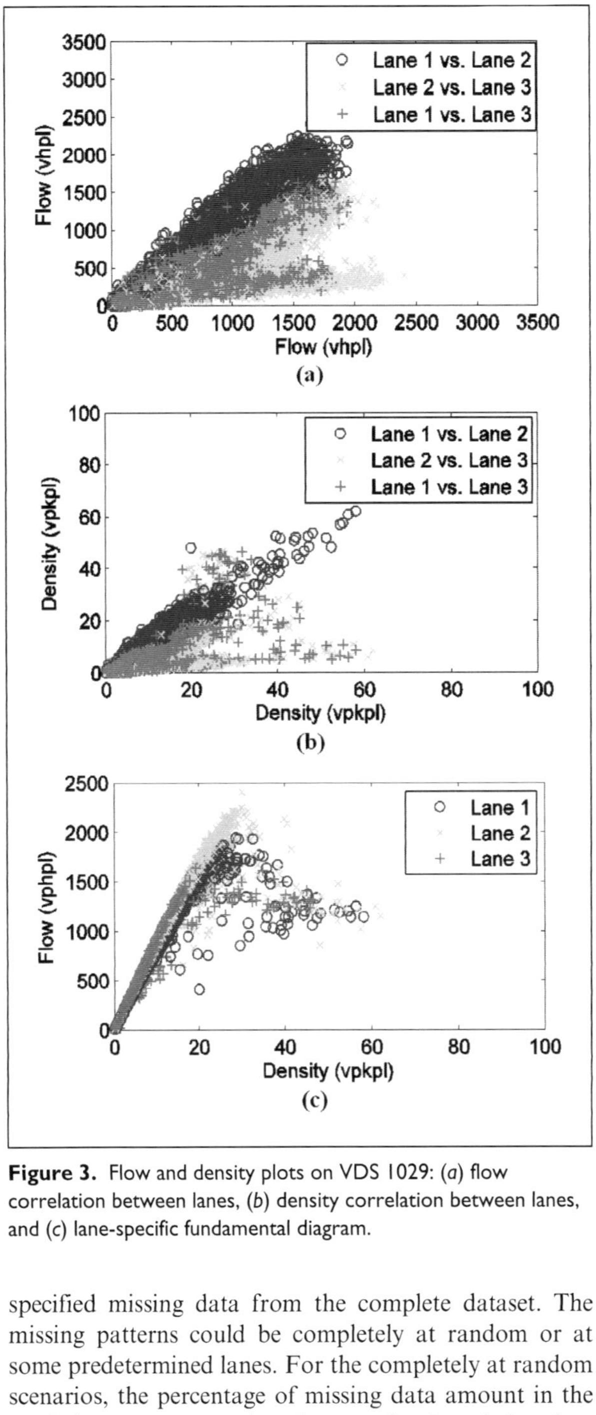

Figure 3. Flow and density plots on VDS 1029: (*a*) flow correlation between lanes, (*b*) density correlation between lanes, and (*c*) lane-specific fundamental diagram.

specified missing data from the complete dataset. The missing patterns could be completely at random or at some predetermined lanes. For the completely at random scenarios, the percentage of missing data amount in the total data amount, or in other words, the missing data ratio, could also be adjusted. To be specific, the software can generate random matrices using prescribed alphabet. In this study, a $T \times N$ matrix, which only contained 0 and 1, was first generated. The percentage of zeros, namely the missing data ratio, was predefined. Using the random missing pattern matrix, the evaluation application removed the specified missing data from the complete dataset. In this way, the dataset with a random missing pattern but with a predefined missing data ratio was obtained.

The study authors assumed that traffic data were complete in the first 5 minutes of each day. This assumption guaranteed online imputation optimization from a lack of sufficient preceding data. Data were queried from the database every minute and screened by the diagnostics algorithm. The diagnostics algorithm generated a data completeness matrix and input it into the imputation module. The imputation module conducted the imputation and filled the imputes into the missing holes. The imputes were treated as true values in the imputation for the next time steps.

Evaluation of Involvement of Temporal Trend. To evaluate the performance of the involvement of temporal trend in the imputation model, this study firstly compared the proposed model with the one that considers only spatial relationships. Model 1 is the proposed model (i.e., Equation [2]). Model 2 (i.e., the following equation) is the generalized form of the model in (*14*), which utilizes only spatial relationships to correct missing data. It is important to note that this part of the evaluation applied offline calibration rather than online calibration, and the missing data pattern was set at a predetermined lane rather than completely random. To be short, static parameter sets were predetermined and input into both models. The data on a predetermined lane were assumed missing all days.

$$x^m_{\text{est},t} = \Delta^m_t x^m_t + (\Omega - \Delta^m_t)(\alpha^m_t \Delta^m_t x^m_t + \gamma^m_t + \varepsilon^m_t) \quad (15)$$

Traffic data from VDS 1029 were used in this part. The models were calibrated with data from the weekdays of the first 2 weeks in May 2015. Then the resulting models were tested with the data from the weekdays of the third week in that month. Table 2 exhibits the coefficients and statistics of the imputation models. Note that some coefficients are not applicable for some model parameters: for both models, $\alpha_{nn'}$ are all not applicable when $n = n'$; and for Model 2, β_n are not applicable. Table 2 indicates that the consideration of both temporal and spatial variables improves the imputation accuracy for all of the scenarios. All of the R^2 values resulting from Model 1 were larger than 0.8. Also, all of the MAPEs and NRMSEs were reduced by Model 1. Most importantly, the most remarkable improvement happened on lane 3. As shown in Table 1, the correlations of flow and density between lane 3 and the other lanes are not significantly obvious compared with that between lane 1 and lane 2. The imputation model relying only on spatial relationships cannot capture the real evolutions on the lane that is not significantly correlated with the others. The coefficient of β_3 in Model 1 also shows that the

Table 2. Coefficients and Statistics of Imputation

Flow imputation

		Model 1			Model 2		
Missing data lane index n		1	2	3	1	2	3
Calibration	α_{1n}	N/A	0.691	0.186	N/A	1.185	1.186
	α_{2n}	0.456	N/A	−0.048	0.677	N/A	−0.551
	α_{3n}	0.139	−0.078	N/A	0.199	−0.161	N/A
	β_n	0.381	0.479	0.934	N/A	N/A	N/A
	γ_n	−16.113	−19.572	−86.968	63.583	115.111	128.013
	R^2 statistic	0.881	0.871	0.848	0.855	0.827	0.330
	F statistic	66309	60880	50102	79361	64390	6659
	p value	0	0	0	0	0	0
Validation	MAPE	0.089	0.089	0.161	0.096	0.101	0.471
	NRMSE	0.109	0.110	0.179	0.115	0.123	0.393

Density Imputation

		Model 1			Model 2		
Missing Data Lane Index		1	2	3	1	2	3
Calibration	α_{1n}	N/A	0.926	0.123	N/A	1.083	0.763
	α_{2n}	0.814	N/A	−0.012	0.819	N/A	−0.393
	α_{3n}	0.163	−0.110	N/A	0.165	−0.112	N/A
	β_n	0.011	0.245	0.933	N/A	N/A	N/A
	γ_n	−0.430	−0.033	−0.955	−0.380	1.620	3.471
	R^2 statistic	0.913	0.922	0.822	0.913	0.905	0.262
	F statistic	94552	106322	41642	141788	128529	4800
	p value	0	0	0	0	0	0
Validation	MAPE	0.098	0.092	0.182	0.098	0.099	0.508
	NRMSE	0.122	0.121	0.475	0.123	0.127	0.628

involvement of temporal trend improves the imputation performance remarkably. Thus, including the temporal trend into the imputation model notably improves imputation performance.

Evaluation of Online Calibration. After confirming above the effectiveness of including the temporal trend in the imputation model, this section clarifies the imputation performance by online imputation and its applicability in traffic control systems. Two types of imputation were implemented: (1) imputation with online calibration; and (2) imputation with offline calibration. According to the procedure in the last section, online calibration generates model parameters dynamically based on preceding traffic data. In the meantime, offline calibration applies static model parameters calibrated from historical data. Traffic data from VDS 1029 and VDS 1031 in May 2015 were used to examine the performance of online calibration. Data in a fixed missing data ratio were removed from the dataset completely randomly. The missing data ratio ranged from 5% to 80%. For each missing data ratio, imputation was run 20 times with different missing data scenarios generated from 20 random matrices. Figure 4 shows one density imputation case at a 20% missing data ratio on VDS 1029 as an example. It plots the imputes of online and offline calibration compared with original complete data. In contrast with offline calibration imputation, online calibration imputation better matches original data. During PM peak hours, sudden increases and decreases occurred in traffic density. Online calibration imputation captured those density increases and decreases in a timely manner. On the contrary, the offline calibration imputation caused unreasonable fluctuations in the imputation results. In this case, the online calibration decreased MAPE from 0.101 to 0.075 and NRSME from 0.1978 to 0.1490, providing better results than the offline calibration.

When imputation performance in a specific case was confirmed, the general performance in various missing data scenarios and missing data ratios was verified. Previous studies reported that missing data ratios vary from 4% to 15% on average (*23*, *24*), while some detectors even lost 20–25% of the traffic data (*24*). Thus, Table 3 gives the detailed MOE values for imputation with online and offline calibration at missing data ratios from 5% to 30%. Data from 15 weekdays were used in the imputation evaluation. For each missing data ratio,

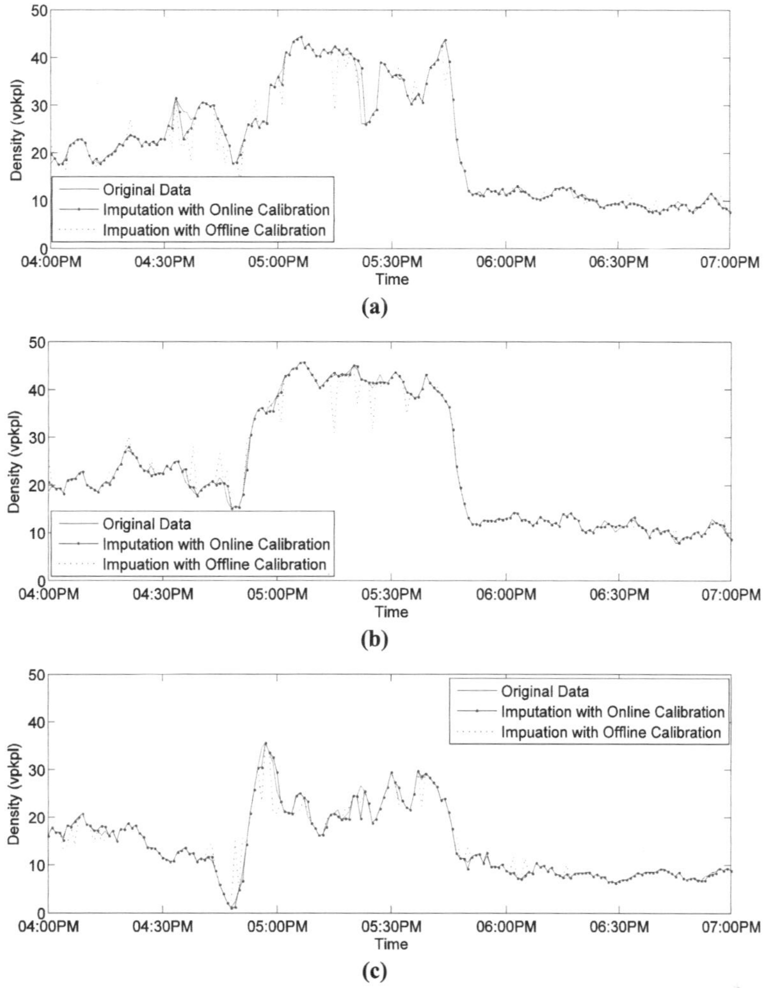

Figure 4. Imputation performance of density at a 20% missing data ratio on VDS 1029 on May 4th, 2015: (*a*) Lane 1, (*b*) Lane 2, and (*c*) Lane 3.

the MOE values were averaged for the 20 missing data scenarios and 15 weekdays, such that there was 300 MOE values for each missing data ratio. As observed in Table 3, the average MOEs for imputation with online calibration are all lower than those for imputation with offline calibration. In addition, Figure 5 graphs the boxplots of the flow imputation MAPEs at VDS 1031 for different missing data ratios. In Figure 5, it can be observed that the MAPEs were not averaged, but each MAPE represented the result from one scenario in one day. Furthermore, the central mark on each box is the median value, and the edges are the 25th and 75th percentiles. The individual data points beyond the whiskers are considered to be outliers. Generally, compared with the imputation with offline calibration, the MAPEs from the imputation with online calibration concentrated more to the median—they varied greater with missing data ratio changes; however, they yielded more outliers. For cases where the missing data ratio was lower than 0.65, the results from the online calibration were distributed closer to the real data. Therefore, the imputation with online calibration enhanced the imputation performance, while the performance sensitivity required further investigations.

Performance Sensitivity Investigation. Figure 5 shows that the sensitivity of missing data ratios differs between imputation with online and offline calibration. Therefore, this study also examined performance sensitivity. Figure 6 exhibits the sensitivities of MOEs by missing data ratios. When the missing data ratio increased, the imputation errors also increased regardless of whether the imputation was equipped with online or offline calibration. The imputation with online calibration was more sensitive to

Table 3. Average Imputation Performance

		Imputation of density				Imputation of flow			
		Online calibration		Offline calibration		Online calibration		Offline calibration	
	Missing data ratio	MAPE	NRSME	MAPE	NRSME	MAPE	NRSME	MAPE	NRSME
VDS 1029	5%	0.063	0.074	0.125	0.149	0.060	0.069	0.105	0.116
	10%	0.065	0.078	0.127	0.155	0.063	0.072	0.106	0.117
	15%	0.069	0.083	0.128	0.159	0.066	0.076	0.107	0.118
	20%	0.072	0.088	0.130	0.162	0.069	0.080	0.108	0.120
	25%	0.076	0.094	0.131	0.165	0.073	0.084	0.109	0.121
	30%	0.080	0.100	0.133	0.170	0.077	0.089	0.110	0.122
VDS 1031	5%	0.066	0.097	0.113	0.171	0.079	0.081	0.129	0.126
	10%	0.069	0.102	0.115	0.177	0.082	0.085	0.131	0.127
	15%	0.072	0.109	0.117	0.187	0.086	0.089	0.131	0.128
	20%	0.076	0.116	0.119	0.196	0.090	0.093	0.132	0.129
	25%	0.080	0.122	0.121	0.205	0.094	0.097	0.134	0.130
	30%	0.084	0.129	0.123	0.214	0.099	0.102	0.134	0.131

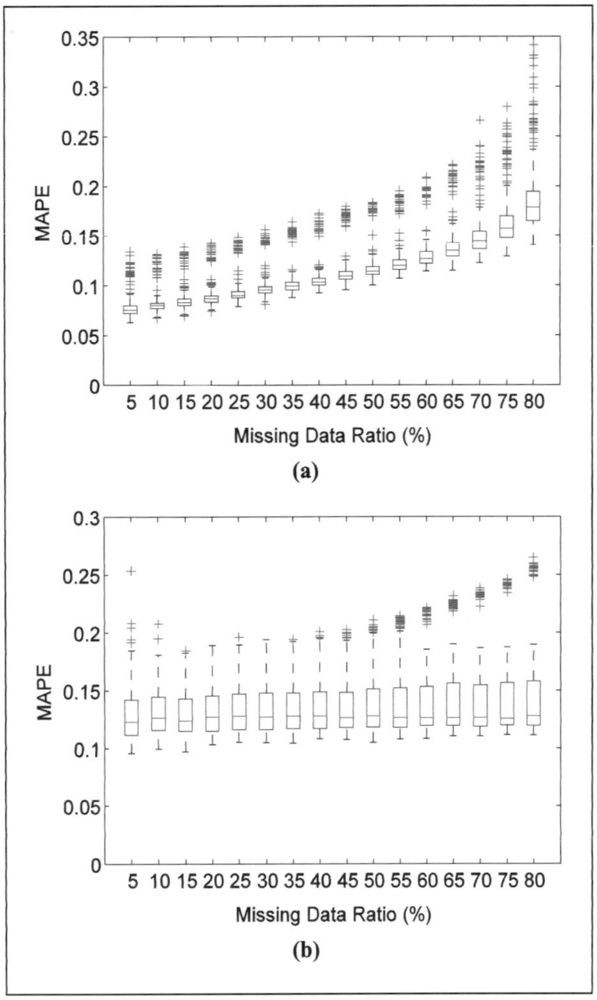

Figure 5. MAPE box plots of flow imputation at VDS 1031 with (*a*) online calibration and (*b*) offline calibration.

missing data ratios—the relationship of MOEs in the imputation with offline calibration and missing data ratios was nearly linear. Meanwhile, MOEs in the imputation with online calibration increased exponentially. At low missing data ratios, imputation with online calibration was observably more accurate than that with offline calibration. However, after the missing data ratio exceeded 60–70%, the offline calibration performed better than the online calibration. The reason is that as the missing data ratio increases, the imputation with online calibration can find less ground truth for reference; but a higher missing data ratio has little impact on the imputation with offline calibration as it is based on historical knowledge.

Thus, offline calibration is more suitable for offline applications with a large missing data ratio, such as average annual daily traffic analysis, whereas online calibration is better for real-time applications, such as real-time traffic control systems. As real-time traffic control systems demand high-quality traffic data, data quality in such systems should be confirmed before being applied. With a low missing data ratio, the imputation with online calibration will outperform offline calibration. However, once the missing data ratio becomes too high to guarantee acceptable imputation accuracy, the control systems should report the data quality issue to operators, and repair will then be necessary. In conclusion, the proposed imputation model with online calibration can correct missing or erroneous data in real time with desirable accuracy and computational simplicity. The proposed method is applicable in real-time traffic control systems.

Conclusions and Future Work

Real-time freeway control systems require high-quality real-time data in implementation; however, traffic data

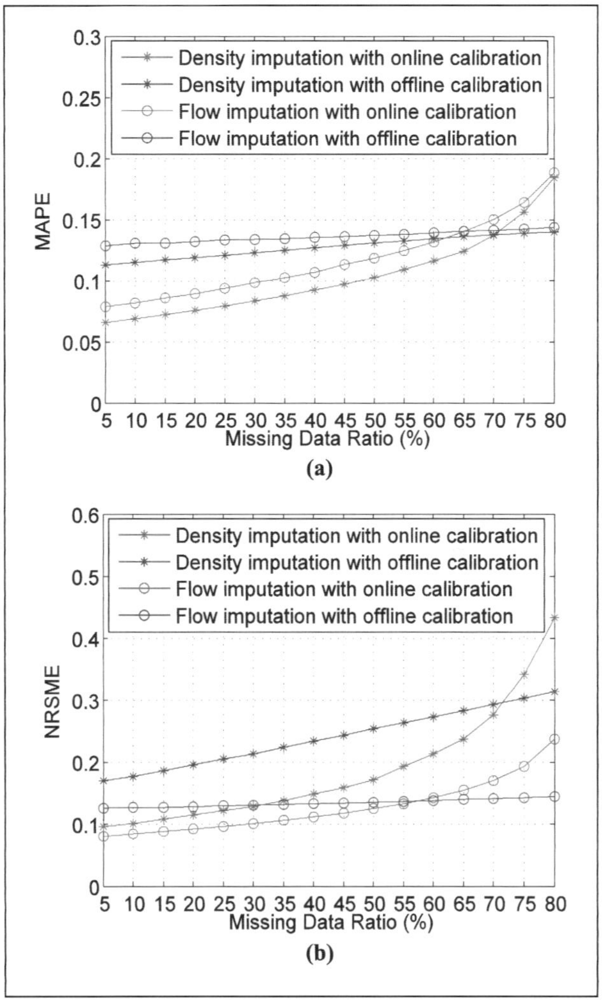

Figure 6. Sensitivity of imputation performance at VDS 1031 with various missing data ratios: (*a*) MAPE and (*b*) NRSME.

are highly prone to experiencing issues related to missing or erroneous data. This paper presented an imputation model with online calibration based on linear regression of temporal trend and spatial relationships. The proposed model was evaluated with authentic traffic data and compared with other methods. The evaluation revealed the following findings:

(a) The correlation between lanes on a station may be low as a result of lane configuration or driver behaviors. The involvement of temporal trend can enhance the imputation accuracy, especially for the lanes that are not significantly correlated with the others.

(b) The best-fit imputation model parameters are time variant and station specific. The online calibration can adapt the model to dynamic variations. Also, when the imputation method is implemented, operators do not need any prior parameter calibration and adjustment because of the online calibration module.

(c) The proposed approach outperforms the imputation with offline calibration when missing data ratios are low. Imputation with online calibration is suitable for real-time freeway control systems, whereas imputation with offline calibration performs better for offline analysis with a high missing data ratio.

Based on the proposed imputation method, future work will make effort to evaluate its imputation performance and implement this method in freeway control systems. This study ignored the weather and incident impact. Future work will evaluate imputation effectiveness in adverse weather and incident conditions. A weather or an incident adjustment factor should be involved in the imputation method. Furthermore, this study generated random missing data scenarios. However, it is still questionable whether the imputation method can still be effective in real-world scenarios. The imputation method will be applied in a freeway control system to test its real-time effectiveness.

Acknowledgments

The authors would like to thank the traffic operation group in the City of Edmonton for providing the data. This research work was jointly supported by the National Natural Science Foundation of China (61703236), China Postdoctoral Science Foundation Funded Project (2017M612275) and Shandong Provincial Natural Science Foundation, China (ZR2017QF014).

Author Contributions

The authors confirm contribution to the paper as follows: study conception and design: Dr. Xu Wang; data collection: Mr. Yuechun Ge; analysis and interpretation of results: Dr. Xu Wang and Dr. Lei Niu; draft manuscript preparation: Dr. Xu Wang, Dr. Yi He and Dr. Tony Z. Qiu. All authors reviewed the results and approved the final version of the manuscript.

References

1. Papageorgiou, M., H. Hadj-Salem, and F. Middelham. ALINEA Local Ramp Metering: Summary of Field Results. *Transportation Research Record: Journal of the Transportation Research Board*, 1997. 1603: 90–98.
2. Papamichail, I., M. Papageorgiou, V. Vong, and J. Gaffney. Heuristic Ramp-Metering Coordination Strategy Implemented at Monash Freeway, Australia. *Transportation Research Record: Journal of the Transportation Research Board*, 2010. 2178: 10–20.
3. Wang, X., M. Seraj, Y. Bie, T. Z. Qiu, and L. Niu. Implementation of Variable Speed Limits: Preliminary Test on

Whitemud Drive, Edmonton, Canada. *Journal of Transportation Engineering*, Vol. 142, No. 12, 2016, p.05016007. https://doi.org/10.1061/(ASCE)TE.1943- 5436.0000895.
4. Smith, B., W. Scherer, and J. Conklin. Exploring Imputation Techniques for Missing Data in Transportation Management Systems. *Transportation Research Record: Journal of the Transportation Research Board*, 2003. 1836: 132–142.
5. Daiheng, N., D. L. John, G. Angshuman, and F. Chunxia. Multiple Imputation Scheme for Overcoming the Missing Values and Variability Issues in Its Data. *Journal of Transportation Engineering*, Vol. 131, No. 12, 2005, pp. 931–938. https://doi.org/10.1061/(ASCE)0733-947X(2005)131:12(931).
6. Li, L., Y. Li, and Z. Li. Efficient Missing Data Imputing for Traffic Flow by Considering Temporal and Spatial Dependence. *Transportation Research Part C: Emerging Technologies*, Vol. 34, 2013, pp. 108–120. https://doi.org/10.1016/j.trc.2013.05.008.
7. Tan, H., Y. Wu, B. Cheng, W. Wang, and B. Ran. Robust Missing Traffic Flow Imputation Considering Nonnegativity and Road Capacity. *Mathematical Problems in Engineering*, Vol. 2014, 2014, pp. 1–8. http://dx.doi.org/10.1155/2014/763469.
8. Tang, J., Y. Wang, S. Zhang, H. Wang, F. Liu, and S. Yu. On Missing Traffic Data Imputation Based on Fuzzy C-Means Method by Considering Spatial–Temporal Correlation. *Transportation Research Record: Journal of the Transportation Research Board*, 2015. 2528: 86–95.
9. Tan, H., G. Feng, J. Feng, W. Wang, Y.-J. Zhang, and F. Li. A Tensor-Based Method for Missing Traffic Data Completion. *Transportation Research Part C: Emerging Technologies*, Vol. 28, 2013, pp. 15–27. https://doi.org/10.1016/j.trc.2012.12.007.
10. Chang, H., D. Park, Y. Lee, and B. Yoon. Multiple Time Period Imputation Technique for Multiple Missing Traffic Variables: Nonparametric Regression Approach. *Canadian Journal of Civil Engineering*, Vol. 39, No. 4, 2012, pp. 448–459. http://dx.doi.org/10.1080/03081060802364505.
11. Abdelgawad, H., T. Abdulazim, B. Abdulhai, A. Hadayeghi, and W. Harrett. Data Imputation and Nested Seasonality Time Series Modelling for Permanent Data Collection Stations: Methodology and Application to Ontario. *Canadian Journal of Civil Engineering*, Vol. 42, No. 5, 2015, pp. 287–302. https://doi.org/10.1139/cjce-2014-0087.
12. Chen, C., J. Kwon, J. Rice, A. Skabardonis, and P. Varaiya. Detecting Errors and Imputing Missing Data for Single-Loop Surveillance Systems. *Transportation Research Record: Journal of the Transportation Research Board*, 2003. 1855: 160–167.
13. Henrickson, K., Y. Zou, and Y. Wang. Flexible and Robust Method for Missing Loop Detector Data Imputation. *Transportation Research Record: Journal of the Transportation Research Board*, 2015. 2527: 29–36.
14. Bie, Y., X. Wang, and T. Z. Qiu. Online Method to Impute Missing Loop Detector Data for Urban Freeway Traffic Control. *Transportation Research Record: Journal of the Transportation Research Board*, 2016. 2593: 37–46.
15. Yin, W., P. Murray-Tuite, and H. Rakha. Imputing Erroneous Data of Single-Station Loop Detectors for Nonincident Conditions: Comparison between Temporal and Spatial Methods. *Journal of Intelligent Transportation Systems*, Vol. 16, No. 3, 2012, pp. 159–176. http://dx.doi.org/10.1080/15472450.2012.694788.
16. Smith, B., and J. Conklin. Use of Local Lane Distribution Patterns to Estimate Missing Data Values from Traffic Monitoring Systems. *Transportation Research Record: Journal of the Transportation Research Board*, 2002. 1811: 50–56.
17. Wang, Y., and M. Papageorgiou. Real-Time Freeway Traffic State Estimation Based on Extended Kalman Filter: A General Approach. *Transportation Research Part B: Methodological*, Vol. 39, No. 2, 2005, pp. 141–167. https://doi.org/10.1016/j.trb.2004.03.003.
18. Papathanasopoulou, V., I. Markou, and C. Antoniou. Online Calibration for Microscopic Traffic Simulation and Dynamic Multi-Step Prediction of Traffic Speed. *Transportation Research Part C: Emerging Technologies*, Vol. 68, 2016, pp. 144–159. https://doi.org/10.1016/j.trc.2016.04.006.
19. Payne, H. J., E. D. Helfenbein, and H. C. Knobel. *Development and Testing of Incident Detection Algorithms*. Report No. FHWA-RD-76-20. Federal Highway Administration, Washington, D.C., 1976.
20. Jacobson, L. N., N. L. Nihan, and J. D. Bender. Detecting Erroneous Loop Detector Data in a Freeway Traffic Management System. *Transportation Research Record: Journal of the Transportation Research Board*, 1990. 1287: 151–166.
21. Antoniou, C., M. Ben-Akiva, and H. Koutsopoulos. Online Calibration of Traffic Prediction Models. *Transportation Research Record: Journal of the Transportation Research Board*, 2005. 1934: 235–245.
22. The MathWorks, Inc. *Matlab Primer*. The MathWorks, Inc., Natick, MA. https://www.mathworks.com/help/pdf_doc/matlab/getstart.pdf. Accessed March, 2018.
23. Al-Deek, H., C. Venkata, and S. R. Chandra. New Algorithms for Filtering and Imputation of Real-Time and Archived Dual-Loop Detector Data in I-4 Data Warehouse. *Transportation Research Record: Journal of the Transportation Research Board*, 2004. 1867: 116–126.
24. Qu, L., L. Li, Y. Zhang, and J. Hu. PPCA-Based Missing Data Imputation for Traffic Flow Volume: A Systematical Approach. *IEEE Transactions on Intelligent Transportation Systems*, Vol. 10, No. 3, 2009, pp. 512–522. https://doi.org/10.1109/TITS.2009.2026312.

The Standing Committee on Highway Traffic Monitoring (ABJ35) peer-reviewed this paper (18-03997).

Article

TRR
JOURNAL OF THE TRANSPORTATION RESEARCH BOARD

Transportation Research Record
2018, Vol. 2672(43) 55–64

DOI: 10.1177/0361198118798979
journals.sagepub.com/home/trr
SAGE

Development of Statewide Annual Average Daily Traffic Estimation Model from Short-Term Counts: A Comparative Study for South Carolina

Sakib Mahmud Khan[1], Sababa Islam[1], MD Zadid Khan[1], Kakan Dey[2], Mashrur Chowdhury[1], Nathan Huynh[3], and Mohammad Torkjazi[3]

Abstract

Annual Average Daily Traffic (AADT) is an important parameter for traffic engineering analysis. Departments of Transportation continually collect traffic count using both permanent count stations (i.e., Automatic Traffic Recorders or ATRs) and temporary short-term count stations. In South Carolina, 87% of the ATRs are located on interstates and arterial highways. For most secondary highways (i.e., collectors and local roads), AADT is estimated based on short-term counts. This paper develops AADT estimation models for different roadway functional classes with two machine learning techniques: Support Vector Regression (SVR) and Artificial Neural Network (ANN). The models predict AADT from short-term counts. The results are first compared against each other, using the 2011 ATR data, to identify the best models. Then, the results of the best models are compared against both the regression-based model and factor-based model. The comparison reveals the superiority of the SVR model for AADT estimation for different roadway functional classes over all other methods. Among models for different roadway functional classes, developed with the 2011 ATR data, the SVR-based models show minimum errors in estimating AADT compared to the ANN-based, regression-based, and factor-based models, depicting the superiority of the SVR-based model for all roadway functional classes over other models in terms of AADT estimation accuracy. SVR models are validated for each roadway functional class using the 2016 ATR data and short-term count data collected by the South Carolina Department of Transportation (SCDOT). The validation results show that the SVR-based AADT estimation models can be used by the SCDOT as a reliable option to predict AADT from the short-term counts.

Annual Average Daily Traffic (AADT) is required in many transportation engineering projects (e.g., roadway design, transportation planning, traffic safety analysis, highway investment decision making, highway maintenance, air quality compliance studies, and travel demand modeling). The ability to accurately estimate AADT using short-term counts is critical for any transportation project that uses AADT as an input variable. For example, AADT is a vital input variable for the Safety Analyst software and the Highway Safety Manual (*1*). Moreover, as part of the traffic-monitoring program, the Federal Highway Administration requires the statewide estimated AADT from every state Department of Transportation (DOT) annually (*2*).

Using permanent traffic count stations or Automatic Traffic Recorders (ATRs), AADT can be directly measured. An ATR captures traffic data 24 hours per day for the entire year using sensors including traditional inductive loops, microwave radar sensors, magnetic counters, and piezoelectric sensors. However, installing ATRs at thousands of traffic count stations is not economically viable. For this reason, ATRs are installed at a handful of strategic locations. To supplement these ATRs, short-term traffic counts (i.e., 24/48-hour counts) are conducted at other locations, and the data collected are used to estimate AADT at these locations. The data collection frequencies at short-term count stations are inconsistent among different states. Whereas short-term

[1]Glenn Department of Civil Engineering, Clemson University, Clemson, SC
[2]Civil and Environmental Engineering, West Virginia University, Morgantown, WV
[3]Department of Civil and Environmental Engineering, University of South Carolina, Columbia, SC

Corresponding Author:
Address correspondence to Sakib Mahmud Khan: sakibk@g.clemson.edu

counts are collected annually in some states, others collect every few years (*3*).

Traditionally, for the locations that do not have an ATR, the AADT is estimated using expansion factors (i.e., seasonal, daily, monthly, growth, and/or axle factors). Many DOTs, such as the South Carolina DOT (SCDOT), use this method to determine AADT from short-term counts. The expansion factors are derived using the continuous traffic volume data collected from the ATRs (*4*). To develop accurate expansion factors, permanent and short-term count stations are combined based on the roadway functional classes and geographical locations (*5*). After grouping, ATR data are used to determine the expansion factors. Data from short-term count stations within the same group are used to determine AADT for those stations by applying these expansion factors. This method of AADT estimation that uses short-term counts is known as the factor-based method, and has some drawbacks. There are no defined guidelines or established standards for allocating the expansion factors from ATR to short-term traffic count stations (*3*). Moreover, the relatively small number of ATRs deployed in the lower roadway functional classes makes it challenging to develop accurate expansion factors for the larger number of short-term count stations on local roads. One solution is to have more ATRs in the lower functional classes, but that has a significant investment requirement. To address this issue, several models are developed in this study based on machine learning. More specifically, the study objective is to develop AADT estimation models that can be used by DOTs to reliably estimate AADT using short-term counts. These models are compared against the factor-based models traditionally used by DOTs to estimate AADT, and regression-based models. The following sections discuss the related studies, method, findings, and research contributions from the developed AADT estimation models.

Literature Review

A brief overview of previous work on AADT estimation methods is presented in this section. Among all methods, regression analysis is widely used for AADT estimation. Xia et al. considered roadways characteristics in AADT estimation in Florida (*6*). Zhao and Chung used GIS to derive land-use and accessibility information for the regression models (*7*). Zhao and Park estimated regression parameters locally (i.e., based on observations near the estimation location) instead of globally and referred to it as the Geographically Weighted Regression (GWR) method (*8*). The comparison showed that GWR is more accurate than the Ordinary Linear Regression. Jiang et al. proposed to use the weighted average of (i) the count from the growth factor-based method, which uses last years' data to predict AADT, and (ii) the traffic count from the current year's image containing traffic information (*9*). Kingan and Westhuis proposed a robust regression method, which minimizes a proportion of the sum of the smallest squared residuals for AADT estimation as the Ordinary Least Square method is susceptible to outliers (*10*). Yang et al. studied variable selection and parameter estimation using various variable groups (*11*). Significant variables can be identified by the Smoothly Clipped Absolute Deviation penalty method. This method can also determine regression coefficients.

Among the machine learning techniques, Artificial Neural Network (ANN) has been used in many studies to evaluate driver behavior, pavement maintenance, vehicle classification, traffic forecasting, and pattern analysis (*12*). A well-trained ANN can model the complex relationship of hourly traffic volume data, find the traffic pattern, and estimate the AADT without grouping the permanent and short-term count stations (*13*). Neural networks have also been used to determine AADT using short-term traffic counts (*3, 13*). Support Vector Regression (SVR) is used for similar applications, specifically for predicting and comparing travel times with the base-line travel time prediction using real-time traffic data. A study by Lin revealed that SVR has greater learning potential than ANN (*14*). However, little research has been conducted using SVR in traffic data analysis (*15*). In the exploration of the potential of SVR for short-term traffic speed predictions, Vanajakshi and Rilett found that when training data are limited, SVR performed better then ANN, although further study is required to determine the advantages and disadvantages of both methods (*15*). Furthermore, in their evaluation of the performance of a modified SVR, which is SVR with Data-Dependent Parameters (SVR-DP), Castro-Neto et al. used 1985–2004 AADT values from Tennessee (*16*). A subsequent comparison of the SVR-DP approach with the popular Holt–Winters exponential smoothing method and the Ordinary Least Squares (OLS) linear regression method showed that the SVR-DP outperformed both (*16*).

Method

This section outlines the five steps followed for AADT estimation model development (i.e., ANN, SVR, and OLS regression) and evaluation. The first three steps include data collection, data processing, and feature selection. Step 4 is the model development step and step 5 is the model evaluation step.

Step 1: Data Collection

ATR Data. The SCDOT maintains a total of 164 ATRs on different functional classes, which include 83 stations on interstates, 59 on arterials, 15 on collectors, and seven

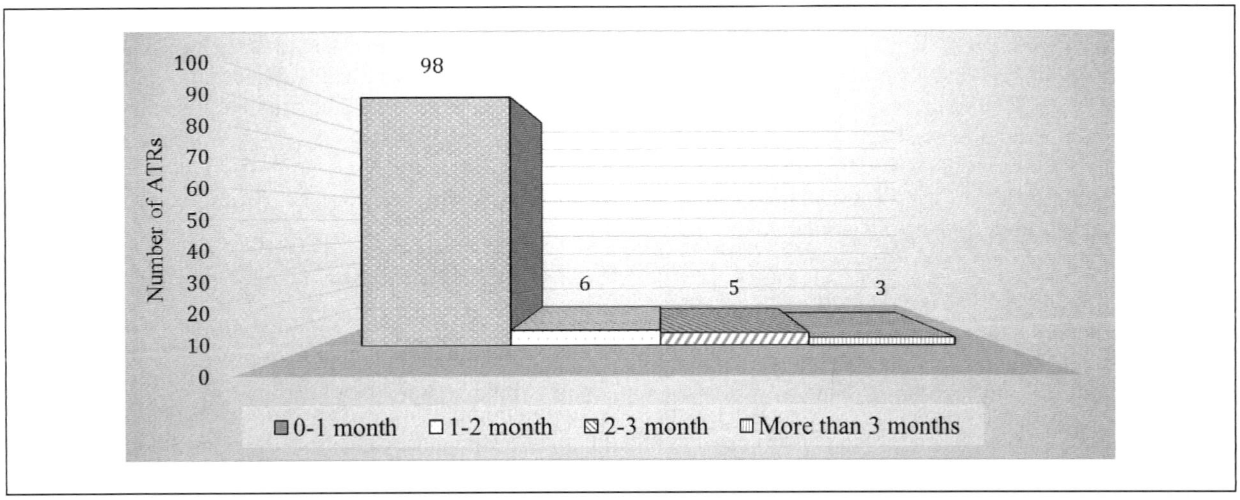

Figure 1. Number of ATRs with missing counts.

on local roads. An interactive web-scraping model, developed in Python using a library called Selenium, is used to collect data from the SCDOT ATR data reporting website. ATRs with more than six months of missing data are not included. Data for a day are not used if any hourly volume data for that day are missing, which can be caused by ATR hardware and/or software malfunctions (*17*). Figure 1 shows the number of ATRs, which was finally selected after excluding ATRs with more than six months of missing data, with the missing counts by the corresponding number of months. Among the 112 ATRs, the figure shows that 98 ATRs had less than one month of missing data, whereas 14 ATRs had missing data for more than one month. Data are collected for multiple years (i.e., for year 2011 and year 2016) to validate the robustness of the models.

Socio-Economic Data. In addition to the ATR data, the authors have collected the following socio-economic information in the zip-code level: (i) income, (ii) employment, (iii) percent below poverty, (iv) number of vehicles, (v) area type (i.e., urban or rural), and (vi) number of housing units from the U.S. Census Bureau data (*18*).

Step 2: Data Preprocessing

After data are collected, they are prepared as the input into the models. The variables are divided into four groups: (i) traffic volume features, (ii) socio-economic features, (iii) roadway characteristics, and (iv) categorical features. For the ANN and SVR models, variables of categories (i), (ii), and (iv) are considered. For the regression models, variables of categories (i), (ii), and (iii) are considered. For the machine learning-based models, two-thirds of the data are used as training data, and one-third of the data are used to test the models.

Traffic Volume Features. To develop the machine learning-based AADT estimation models, the authors have followed the study conducted by Sharma et al., in which ANN is used to develop an AADT estimation model for Minnesota (*3*). However, rather than using 48 hourly volume factors in the Minnesota model, the authors have used 24 hourly volume factors in this study. The equation for developing the hourly volume factor is expressed below:

$$\begin{aligned} &\textit{Hourly volume factor} \\ &= \frac{\textit{Traffic volume for an hour } (\textit{e.g., traffic volume for } 7a.m. - 8a.m. \textit{ on monday of January}, 2011)}{\textit{Sum of } 24 \textit{ hourly volume } (\textit{e.g., } 24 - \textit{hour traffic volume on monday of January}, 2011)} \end{aligned} \quad (1)$$

For regression-based models, hourly volume data are directly used to estimate AADT, rather than converting them to hourly factors.

Socio-Economic Feature. The zip-code level data from the U.S. Census are processed using ArcGIS. The geographic map of the traffic counts and roadway characteristics are integrated with the socio-economic data at the zip-code level.

Roadway Characteristics. Roadway characteristics are the most widely used variables in AADT estimation. In this research, only one roadway characteristic is considered: the functional classification. Roadways from a higher functional class with a higher number of lanes always have the higher traffic volume. According to ATR stations' roadway functional classification, three groups of functional classes are considered for the model development step: (i) interstates and expressways, (ii) principal and minor arterials, and (iii) all-ATR models.

Categorical Features. Most AADT estimation models only use hourly volume data (continuous features/variables) (*3, 13*); no categorical features/variables are used along with the continuous hourly volume data. In this study, however, the authors have developed models with continuous (hourly volume) and categorical features. The categorical features that have been considered are: (i) day-of-week and (ii) month-of-year. The reason for considering these two variables is that they have a significant impact on traffic volumes. Binary variables are used for quantifying these categorical features. For example, to develop the day-of-week variables, one feature is used for each day of the week for a total of seven features. If a data point is for Monday, then the Monday feature is assigned the value 1, and the features for the other days of the week are assigned 0. A similar method is used to develop the month-of-year categorical features.

AADT Factors. The target feature used in this study is a factor of the actual AADT called the AADT factor. For each day, the AADT factors are calculated using Equation 2:

$$\text{AADT factor for a specific day} = \frac{AADT}{Sum\ of\ 24\ hourly\ volume\ of\ that\ day} \quad (2)$$

AADT is estimated using 24 hourly volume data for each day. For each ATR, the AADT is computed by calculating a simple average mean of the estimated AADT from each day of the year as mentioned in the traffic monitoring guide (*9*).

Step 3: Feature Selection

Feature selection for ANN and SVR models is performed to reduce the use of irrelevant variables in developing predictive models, and to improve accuracy of the models (*10*). The sequential feature selection method is used to identify the desired features (*19*). It is a simple Greedy Search method starting with an empty set of features and sequentially adding the most impactful feature in each step until the desired result is achieved by satisfying a single criterion or multiple criteria (*11, 17*). In this study, the combination of features with the least residual sum of square error for predicting target values and features are chosen. The sequential feature selection method is applied for the 24 hourly volume data, and 20 hourly volume data are considered finally. For the regression models, most significant features to estimate AADT are identified using the stepwise regression method. No independent variables were found to be correlated.

Step 4: Model Development

Once the hourly volume features are selected, ANN and SVR models are developed using a combination of the hourly volume features with the categorical features and socio-economic features, where the output is the AADT factor. For the regression models, the dependent variable is the AADT. The hourly volume, socio-economic factors, and roadway characteristics are the independent variables. At first, three major models were developed using ANN and SVR: (i) the interstate and expressway model, (ii) the principal and minor arterial model, and (iii) the general model or all-ATR model which can be applied to any roadway functional class. The input data set is filtered accordingly, as the models are functional class specific. Depending on the combinations of hourly volume factors, socio-economic, and categorical features, nine alternatives are created for each functional class model. This is done to determine which combination of the features produces the best model. Figure 2 shows the combination of features for each alternative. The interstate and expressway model, and the principal and minor arterial model are developed using the permanent traffic count stations available for the specific roadway functional class.

The all-ATR model uses data from all available ATRs. The motivation for developing this model is to create a general model that is not roadway functional class specific, and is applicable for estimating AADT on any functional class from the short-term counts. The all-ATR model has only four alternatives instead of nine, alternative 2, 3, 4, and 5. This is because the alternatives consisting of socio-economic features are not used, since those data are not available for all ATRs. The following sections discuss model development using ANN, SVR, and OLS regression.

Artificial Neural Network. In this study, the authors have used a multilayered, feed-forward, back-propagation neural network model for supervised learning. The developed neural network model has three layers, which are the input, hidden, and output layer. This ANN model is a feed-forward network as it feeds the output of one layer to the next layer. A tan-sigmoid transfer function is used to calculate the output from each neuron (*20*). One of the remarkable characteristics of a back-propagation neural

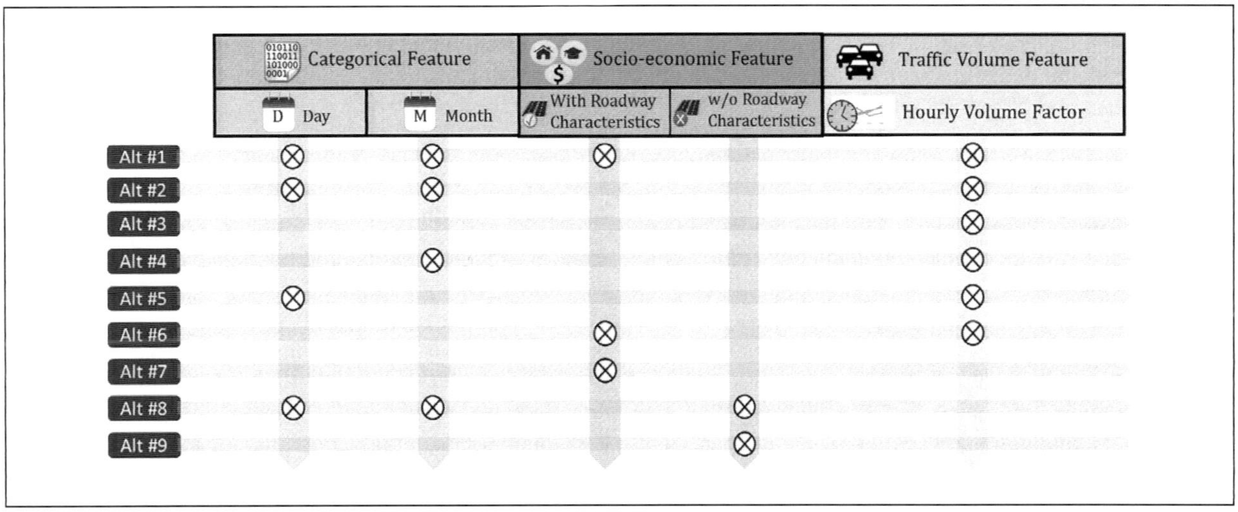

Figure 2. Model alternatives to estimate AADT.

network is that it can propagate the effects of error backward through the network and update weight after every training case (*21*); therefore this algorithm is chosen for estimating AADT in this study. The training algorithm selected is the Levenberg–Marquardt, which reduces the computation complexities of prediction problems (*22*).

In this study, the authors have built different ANN models with a different number of hidden neurons and chose the number of neurons which provides the least root mean square error (RMSE) for the models. The number of hidden neurons used in this study varies for different models. The neural network models are developed using the MATLAB library function NNtool.

Support Vector Regression. SVR is used for nonlinear regression, and it maps the training set onto a kernel-induced, higher dimensional feature space. In this study, a SVR algorithm with radial basis function kernel is chosen and implemented using the MATLAB LIBSVM library tool (*23*). The cost function (C) is used to determine the tradeoff between the complexity of the model and degree of deviation from ε (epsilon) that can be tolerated. In order to fit the training data, the ε parameter is used (*23*). The model relies on a single subset of the training samples; the cost function ignores the training samples within the ε tube (i.e., a certain threshold distance from the prediction). The initial value of C and γ (gamma) are determined based on the grid search method with a 5-fold cross-validation. Cross-validation is conducted to reduce the bias of a training dataset on the model parameters. The parameters that yield the least error on AADT prediction are found using trial and error method. C and γ values are used as powers of two, where the range of powers for C values are −3 to 15, and the range of powers for γ values are −15 to 3, as suggested in Chang and Lin (*23*). The value of the parameters varies from model to model with the change in training data.

Regression. The OLS regression method is used to develop the model. OLS is a method used to determine the unknown parameter in a linear regression model. The dependent variable in the regression model is AADT. Hourly volume data, roadway characteristics, and socio-economic characteristics are used as independent variables for the models.

Step 5: Model Evaluation

Once the alternatives for each machine learning-based model have been developed, the results are compared against the ground truth AADT from the ATR data to calculate the model accuracy. Based on the accuracy of the alternatives, a final alternative for each functional class and a general model are selected as the best models. For regression, three models are developed, one of which is the general model.

The performance of the alternatives is assessed based on the RMSE and mean absolute percentage error (MAPE). The formula used for RMSE and MAPE are shown below as Equations 3 and 4:

$$\mathrm{RMSE} = \sqrt{\frac{\sum_{i=1}^{n} (Y - yi)^2}{n}} \tag{3}$$

$$\mathrm{MAPE} = \frac{1}{n}\sum \frac{|Y - yi|}{Y} * 100 \tag{4}$$

where Y = actual value, y_i = predicted value for ith observation, and n = number of observations.

The best alternative for each model is selected based on RMSE and MAPE. Then, the best alternatives for each functional class are compared against the factor-based model and regression-based model.

After that, the robustness and effectiveness of the selected models are validated in two steps. First, the models are validated using a different input dataset, such as ATR data from a different year. The whole model development process is followed again, but with the ATR data from a different year. This step is very important to prove the transferability of the models. If the performance of the models is satisfactory, then the AADT estimation models can be used for any future year.

Second, the authors have used short-term counts as input to the models and predict the AADT for that roadway. The predicted AADT is compared with the AADT calculated from the nearby ATR data, which represents the ground truth data in this case. If the prediction error is reasonable, then it can be concluded that the models perform well in predicting AADT from short-term counts, and they can be used by DOTs for statewide estimation of AADT from short-term counts.

Analysis and Results

As a case study, 2011 is chosen as the year for all data collection for model development, and 2016 data are used for machine learning-based model validation. From the 2011 ATR data, a total of 112 ATRs are used in the model development. Other ATRs had insufficient data (e.g., missing more than six months of data) for accurate AADT estimation. Socio-economic data are obtained from the 2011 census data.

AADT Estimation with Machine Learning Techniques

Using the input data from 2011, model alternatives are developed. Then, the results from ANN and SVR models are summarized to determine which alternatives perform better based on two performance measures, RMSE and MAPE. The alternatives with the least error are identified and then compared against the regression-based model and factor-based model. After this comparison, the best alternatives for ANN and SVR with the lowest RMSE and MAPE are carried forward to the evaluation step, where the alternatives are validated using a new dataset from the year 2016. Table 1 summarizes the results of the ANN and SVR models.

Model 1: Interstate and Expressway Models. The RMSE and MAPE values for different alternatives of the Interstate and Expressway functional class are shown in Table 1. Each alternative consists of different combinations of training features, such as traffic volume factors, socio-economic variables, and other categorical features (i.e., day-of-week, month-of-year). The performances of the

Table 1. ANN and SVR Alternative Models' Performance

		Performance in terms of RMSE and MAPE			
		ANN		SVR	
Model	Model alternative	RMSE	MAPE (%)	RMSE	MAPE (%)
Interstate and expressway model	Alternative 1	0.37	17.6	0.37	16.7
	Alternative 2	0.25	11.9	0.22	11.3
	Alternative 3	0.26	13.8	0.23	12.7
	Alternative 4 (Monday)	0.27	12.5	0.23	11.5
	Alternative 5 (January)	0.52	17.8	0.26	11.7
	Alternative 6	0.48	27.0	0.37	16.6
	Alternative 7	0.44	25.4	0.37	16.7
	Alternative 8	0.38	19.8	0.38	19.8
	Alternative 9	0.37	19.5	0.37	16.7
Principal and minor arterial model	Alternative 1	1.19	51.5	0.34	17.3
	Alternative 2	0.29	17.4	0.29	13.7
	Alternative 3	0.32	18.7	0.31	14.1
	Alternative 4 (Monday)	0.46	12.5	0.32	11.5
	Alternative 5 (January)	0.53	17.8	0.35	11.7
	Alternative 6	2.09	85.6	0.35	16.5
	Alternative 7	2.15	85.6	0.35	17.6
	Alternative 8	0.41	21.7	0.34	16.5
	Alternative 9	2.60	124.4	0.35	17.2
All-ATR model	Alternative 2	0.36	17.4	0.35	15.1
	Alternative 3	0.32	15.9	0.31	13.7
	Alternative 4	0.32	15.6	0.46	18.4
	Alternative 5	0.31	14.8	0.33	13.6

Table 2. Summary of Regression Models

Model	Variable	R-squared value
Interstate and expressway model	• Traffic volume features: Count for 1–4, 6–7, 9–14, 16, 18–24 hour count • Roadway characteristics: Functional class • Socio-economic features: Area type, Income, Employment, Person below poverty, Vehicles, Housing unit	0.884
Principal and minor arterial model	• Traffic volume features: Count for 1–5, 7–8, and 10–22 hour count • Roadway characteristics: Functional class • Socio-economic features: Area type, Income, Employment, Person below poverty, Vehicles, Housing unit	0.973
All-ATR	• Traffic volume features: 1–5, 7–8, 10–16,18–20 and 22 hour count • Roadway characteristics: Functional class • Socio-economic features: Area type, Income, Person below poverty, Vehicles	0.924

SVR alternatives are better than those of ANN alternatives. The best alternative found from this comparison is the SVR alternative 2, which has day, month, and hourly volume factors as input features, with an RMSE of 0.22 and MAPE of 11.3%. A comparison of the alternative errors shows that the addition of socio-economic features lowers the model performance.

Model 2: Principal and Minor Arterial Model. The RMSE and MAPE of the different SVR and ANN alternatives for principal and minor arterials are shown in Table 1. The results are very similar to Model 1 results. For the principal and minor arterial functional class of roadways, SVR alternatives perform better than ANN alternatives. Moreover, the addition of the socio-economic variables lowers the performance of the models. The SVR alternative 2 has the lowest RMSE (0.29).

Model 3: All-ATR Model. The RMSE and MAPE of different SVR and ANN alternatives for all 112 permanent count stations on highways in South Carolina are summarized in Table 1. The training features used for developing these alternatives are the hourly volume factors, month of year, and day of week. The socio-economic data are not available for all permanent count locations; therefore, not all alternatives are developed for this model. A comparison of the results for the different alternatives shows that SVR alternative 3 produces the lowest RMSE (0.31). Alternative 3 uses only the hourly volume data for all available permanent count stations as input (no other variables are used). This general all-ATR model provides a reasonable estimate of AADT considering the variability of the training dataset.

AADT Estimation with Regression Models

Three regression models are developed, one for expressways and interstates, one for principal and minor arterials, and one for all functional classes. The variables and R-squared values of these models are shown in Table 2.

Comparison between SVR and Regression-Based AADT Estimation Model

To compare the effectiveness of the SVR models, the best alternative for SVR is compared against the results of the traditional regression-based models. For this comparison, the same ATRs that are used for testing ANN and SVR-based AADT estimation models are used. In total, estimated AADT from 30 ATRs (with 351 days for each ATR) are compared. The MAPE (%) is calculated for all three models (i.e., interstate, arterial, and all-ATR model) to evaluate the performance of SVR and regression. The comparison results are presented in Figure 3. Based on this comparison, it can be concluded that the SVR alternative provides more accurate AADT estimation for different roadway functional classes.

Comparison between SVR and SCDOT AADT Estimation (Factor) Method

This study focuses on finding the efficacy of the models developed using the machine learning techniques over the traditional factor-based method currently used by SCDOT. In the traditional factor-based method for estimating AADT, SCDOT uses two types of factors: (1) seasonal or monthly factors, and (2) axle correlation factors. These factors are gathered for each of the roadway functional classes. The short-term counts conducted in these functional classes are multiplied with these functional class-specific factors to estimate AADT. To compare the AADTs estimated by SVR and factor-based methods, the same ATRs that are used for testing ANN and SVR-based AADT estimation models are used. In total, estimated AADTs from 30 ATRs (with 351 days for each ATR) are compared. In the SVR-based models,

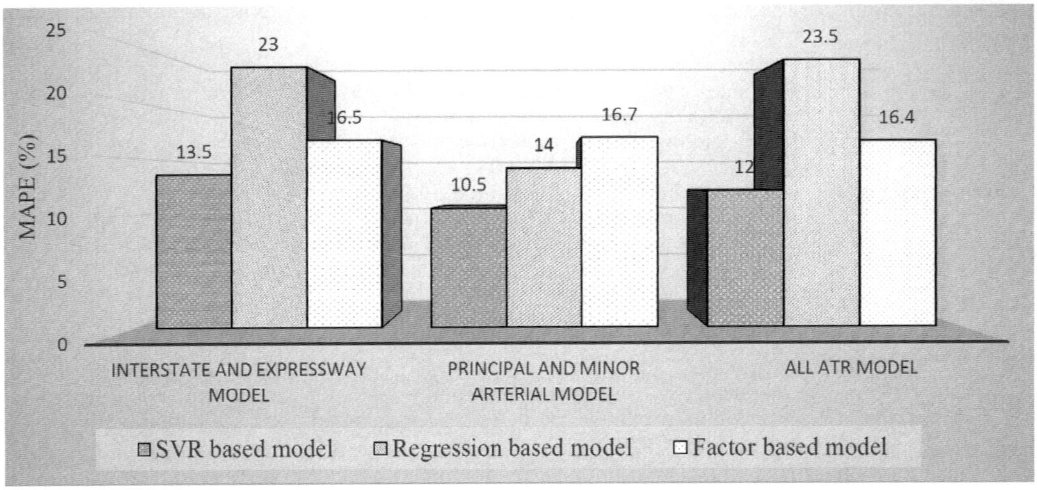

Figure 3. MAPE values for AADT estimation models.

Table 3. SVR Model Validation using ATR Data and Short-Term Count Data

Validation data	Model	RMSE	MAPE (%)
2016 ATR data	All-ATR model (Alternative 2)	0.53	11.9
	All-ATR model (Alternative 3)	0.55	12.7
	Interstate and expressway model (Alternative 2)	0.61	12.8
	Principal and minor arterial model (Alternative 2)	0.31	11.7
Short-term count (5 locations)	All-ATR model (Alternative 2)	0.11	10.2%
	Interstate and expressway model (Alternative 2)	0.09	9.5%

the estimated AADT factors are multiplied by the sum of 24 hourly volumes to calculate the AADT. To estimate AADT using the factor-based method, the sum of 24 hourly volume for the selected day is multiplied by the monthly factor and axle correction factor. In this section, the AADT values are compared for the interstate and expressway model, principal and minor arterial model, and all-ATR models. The MAPE values for the three models are presented in Figure 3. It can be observed that SVR yields a better MAPE value for the principal and minor arterial models than the interstate and expressway model. For the principal and minor arterial model, the MAPE value of the SVR model is 10.5%, which is lower than the MAPE value of the traditional factor-based method (16.7%).

SVR Model Validation

The SVR models are superior for all functional classes. In this section, SVR-based models are validated using two methods. First, the AADT estimation model is developed with the 2016 ATR data, instead of the 2011 ATR data. The only input features required for alternative 2 are 24 hourly volumes and categorical features. The 2016 ATR data are collected from all 164 ATRs. Model alternative 2 is trained with two-thirds of the 2016 ATR data, and tested with one-third of the 2016 ATR data. Alternative 3 is also developed for all-ATR models, as this was the best alternative for all-ATR models using the 2011 dataset. The results are shown in Table 3. Alternative 2 performs better than alternative 3 for the all-ATR model. This represents a change from the 2011 data. As an example, the RMSE values for 2011 and 2016 for alternative 2 of the principal and minor arterial models are 0.29 and 0.31, respectively. The MAPE values for 2011 and 2016 for alternative 2 of the principal and minor arterial models are 13.7% and 11.7%, respectively. These results indicate that the models are robust and stable despite them being developed using datasets from two different years. Based on this validation, it can be concluded that alternative 2 is the best alternative for all three models (i.e., interstate and expressway model, principal and minor arterial model, and the all-ATR model).

The second method used to validate the developed SVR models is to use short-term count data collected from five locations near existing ATRs on interstates in 2016. The ATR IDs of those locations are 98, 145, 102, 50, and 110. The authors have used these data as input to alternative 2 for both interstate and expressway model and the all-ATR model, as these are the models that can be used for interstates. The training dataset for SVR includes all ATRs except for the five ATRs closest to the short-term count locations. The model validation results are shown in Table 3. The interstate and expressway model performs slightly better than the all-ATR model in terms of RMSE and MAPE.

Contributions of this Research

This study introduces several new dimensions to the literature on AADT estimation. The comparisons made between the machine learning-based models, regression-based models, and factor-based models (the factor-based model is currently used by SCDOT) is the first such evaluation. Furthermore, this study investigated multiple roadway functional classes separately, and subsequently developed models and identified the best model for each functional class. Lastly, this is the first study to develop a practical implementation framework for more reliable, statewide AADT estimation models for the state of South Carolina.

Conclusions

In this study, the AADT estimation models are developed to estimate AADT using short-term counts on different roadway functional classes in South Carolina. The AADT estimation models are developed using two machine learning-based models (i.e., ANN, SVR) and a OLS regression-based model. The study analysis indicates that the SVR model outperforms all other models. The best SVR model uses hourly volume data, as well as the day-of-week and the month-of-year categorical features. Adding other features such as socio-economic factors lowers the model performance. This finding suggests that the AADT of a location depends primarily on the traffic patterns (i.e., day-of-week, month-of-year). These models can be incorporated into an AADT estimation software that can be used by DOTs to predict AADT using short-term traffic counts.

Acknowledgments

This work is funded by the South Carolina Department of Transportation.

Author Contributions

The authors confirm contribution to the paper as follows: study conception and design: MC, SI, SMK; data collection: SMK, SI, MDZK, KD, MC, NH, MT; analysis and interpretation of results: SMK, SI, MDZK, KD, MC, NH, MT; draft manuscript preparation: SMK, SI, MDZK, KD, MC, NH, MT. All authors reviewed the results and approved the final version of the manuscript.

References

1. Harwood, D. W., D. J. Torbic, K. R. Richard, and M. M. Meyer. *Safety Analyst: Software Tools for Safety Management of Specific Highway Sites*. FHWA-HRT-10-063. Midwest Research Institute, Kansas City, MO, 2010.
2. Traffic Monitoring Guide. FHWA -PL -95 -031. https://www.fhwa.dot.gov/ohim/tmgbook.pdf. Accessed July 31, 2016.
3. Sharma, S., P. Lingras, F. Xu, and G. Liu. Neural Networks as Alternative to Traditional Factor Approach of Annual Average Daily Traffic Estimation from Traffic Counts. *Transportation Research Record: Journal of the Transportation Research Board*, 1999. 1660: 24–31.
4. Garber, N. J., and L. A. Hoel. *Traffic and Highway Engineering*. Cengage Learning, Toronto, ON, Canada, 2014.
5. Schneider, W. H., and I. Tsapakis. Review of Traffic Monitoring Factor Groupings and the Determination of Seasonal Adjustment Factors for Cars and Trucks. FHWA/OH-2009/10A. University of Akron, Department of Civil Engineering, Akron, OH, 2009.
6. Xia, Q., F. Zhao, Z. Chen, L. Shen, and D. Ospina. Estimation of Annual Average Daily Traffic for Nonstate Roads in a Florida County. *Transportation Research Record: Journal of the Transportation Research Board*, 1999. 1660: 32–40.
7. Zhao, F., and S. Chung. Estimation of Annual Average Daily Traffic in a Florida County using GIS and Regression. *Transportation Research Record: Journal of the Transportation Research Board*, 2001. 1769: 113–122.
8. Zhao, F., and N. Park. Using Geographically Weighted Regression Models to Estimate Annual Average Daily Traffic. *Transportation Research Record: Journal of the Transportation Research Board*, 2004. 1879: 99–107.
9. Jiang, Z., M. R. McCord, and P. K. Goel. Improved AADT Estimation by Combining Information in Image- and Ground-Based Traffic Data. *Journal of Transportation Engineering*, Vol. 132, No. 7, 2006, pp. 523–530.
10. Kingan, R., and T. Westhuis. Robust Regression Methods for Traffic Growth Forecasting. *Transportation Research Record: Journal of the Transportation Research Board*, 2006. 1957: 51–55.
11. Yang, B., S. Wang, and Y. Bao. Efficient Local AADT Estimation via SCAD Variable Selection Based on Regression Models. *Proc., 2011 Chinese Control and Decision Conference (CCDC)*, Mianyang, China, 2011, pp. 1898–1902.
12. Himanen, V., P. Nijkamp, and A. Reggiani. *Neural Networks in Transport Applications*. Ashgate Publishing, Farnham, UK, 1998.
13. Sharma, S., P. Lingras, F. Xu, and P. Kilburn. Application of Neural Networks to Estimate AADT on Low-Volume Roads. *Journal of Transportation Engineering*, Vol. 127, No. 5, 2001, pp. 426–432.

14. Lin, W. A. *Case Study on Support Vector Machine versus Artificial Neural Networks*. University of Pittsburgh, Pittsburgh, PA, 2004.
15. Vanajakshi, L., and L. R. Rilett. A Comparison of the Performance of Artificial Neural Networks and Support Vector Machines for the Prediction of Traffic Speed. *Proc., IEEE Intelligent Vehicles Symposium*, Parma, Italy, 2004, pp. 194–199.
16. Castro-Neto, M., Y. Jeong, M. K. Jeong, and L. D. Han. AADT Prediction using Support Vector Regression with Data-Dependent Parameters. *Expert Systems with Applications*, Vol. 36, 2009, pp. 2979–2986.
17. Qu, L., L. Li, Y. Zhang, and J. Hu. PPCA-Based Missing Data Imputation for Traffic Flow Volume: A Systematical Approach. *IEEE Transactions on Intelligent Transportation Systems*, Vol. 10, No. 3, 2009, pp. 512–522.
18. U.S. Census Bureau; generated by Jane Smith; using *American FactFinder*. http://factfinder2.census.gov. Accessed January 24, 2016.
19. Aha, D. W., and R. L. Bankert. *A Comparative Evaluation of Sequential Feature Selection Algorithms. Learning from Data*. Springer, New York, NY, 1996, pp. 199–206.
20. Islam, S. *Estimation of Annual Average Daily Traffic (AADT) and Missing Hourly Volume using Artificial Intelligence*, Masters Thesis, Clemson University, Clemson, SC, 2016. https://tigerprints.clemson.edu/cgi/viewcontent.cgi?referer=https://scholar.google.com/&httpsredir=1&article=3568&context=all_theses. Accessed March 10, 2018.
21. David Leverington, 2009. http://www.webpages.ttu.edu/dleverin/neural_network/neural_networks.html. Accessed July 30, 2016.
22. Demuth, H., M. Beale, and M. Hagan. Neural Network Toolbox™ 6. *User Guide*, Vol. 10, 2008, p. 11.
23. Chang, C. C, and C. J. Lin. LIBSVM: A Library for Support Vector Machines. *ACM Transactions on Intelligent Systems and Technology (TIST)*, Vol. 2, No. 3, 2011, p. 27.

The Standing Committee on Highway Traffic Monitoring (ABJ35) peer-reviewed this paper (18-06367).

The contents of this report reflect the views of the authors who are responsible for the facts and the accuracy of the presented data. The contents do not reflect the official views of SCDOT or FHWA. This report does not constitute a standard, specification, or regulation.

Article

TRR
JOURNAL OF THE TRANSPORTATION RESEARCH BOARD

Transportation Research Record
2018, Vol. 2672(43) 65–74

DOI: 10.1177/0361198118788188
journals.sagepub.com/home/trr
SAGE

Incorporating the Effect of Special Events into Continuous Count Site Selection for Pedestrian Traffic

Caleb Olfert[1], Rob Poapst[2], and Jeannette Montufar[2]

Abstract

This paper presents results from pedestrian monitoring research conducted in a dense urban environment in Winnipeg, Canada. Pedestrian counts were conducted in downtown Winnipeg using infrared pedestrian counters. Count sites were assigned to traffic pattern groups (TPGs) based on their response to special events occurring in the study area. Once these groups were established, eight continuous count sites were installed to initiate an ongoing pedestrian traffic monitoring program for the city. Traffic monitoring efforts have primarily focused on motorized travel. As more jurisdictions prioritize active transportation, addressing the need for network-level pedestrian data is essential to optimize engineering decisions. The first step to developing any system-wide traffic monitoring program is to define TPGs. These groups enable the spatial variation of short-duration counts to be adjusted to annual statistics by the temporal variation of similarly behaving continuous counts. Short-duration count sites were characterized by daily and hourly trends consistent with existing pedestrian traffic monitoring practices. Recognizing the influence of large evening events on pedestrian traffic, a metric was developed called the evening proportion ratio (EPR) to quantify the effect of special events. Based on the spatial distribution of EPR values, two TPGs were developed for downtown Winnipeg. These are the "urban utilitarian" and "urban utilitarian – event" groups. These groups were used to select continuous count locations for ongoing pedestrian traffic data collection. The importance of this research lies in its future applicability to other jurisdictions in developing a standard approach for urban transportation authorities to strategically implement pedestrian traffic monitoring programs.

In the wake of an increasing number of urban jurisdictions moving to prioritize active travel modes, research into pedestrian and cyclist traffic data collection has progressed beyond its infancy to a point where cities have begun network-wide active transportation monitoring programs. This trend, motivated by shifting societal values toward sustainability, public health, and urban vibrancy, requires accurate and accessible data to inform investment decisions and network improvements. Most pedestrian traffic data collection has been conducted in one of three ways: 1) short-duration manual counts performed by cycling and walking organizations, 2) continuous automated counts performed for academic research, or 3) jurisdictionally motivated ad hoc inclusion of active modes as a part of an existing motorized traffic monitoring program. This research contributes to developing a standard approach for urban transportation authorities to strategically implement pedestrian traffic monitoring programs. This was achieved by developing a new way to characterize pedestrian traffic in highly urbanized areas that frequently host large events. Characterization of pedestrian traffic enables the development of traffic pattern groups (TPGs) which aid jurisdictions in effectively selecting continuous count sites.

To date, most traffic monitoring efforts have focused on pedestrians and cyclists traveling on multi-use paths. Little research has been done on pedestrian travel in dense urban environments with complex travel networks. This paper presents a methodology for assessing pedestrian activity in highly urbanized environments for recurring and non-recurring traffic variation. This paper is based on research conducted over a 4-month period in Winnipeg, Canada, using infrared detectors for pedestrian traffic monitoring.

Following the precedent set by motorized traffic monitoring practices, an effective approach to estimate

[1]Department of Civil Engineering, University of Manitoba, MB, Canada
[2]MORR Transportation Consulting, MB, Canada

Corresponding Author:
Address correspondence to Caleb Olfert: colfert@dillon.ca

network-wide annual average traffic statistics is through a modest number of continuous counts supplemented by a large number of rotating short-duration coverage counts. Continuous counts placed at strategic locations can be used to assess temporal traffic variation, whereas coverage counts provide insight into spatial variation across a network during the count period. Used together, the spatially diverse data collected by coverage counts may be adjusted by the known temporal traffic variation of similarly behaving continuous count sites. This allows annual average pedestrian traffic statistics to be estimated wherever these coverage counts are conducted.

In 2015, the city of Winnipeg and the University of Manitoba began to develop a pedestrian traffic monitoring program in the downtown area. The first objective was to select eight continuous count locations that would monitor pedestrian traffic in winter and summer months. Eight were selected due to limited availability of resources. It was observed that pedestrian movement in downtown Winnipeg was dictated by two concurrent patterns: 1) a temporally recurring traffic pattern dominated by commuter work schedules, and 2) a non-recurring traffic pattern influenced by special events hosted in the downtown arena. Over the 4-month study period, data was collected to better understand this phenomenon and develop a standard approach for selecting continuous pedestrian count sites. This method was designed to characterize pedestrian activity in highly urbanized networks amidst the frequent influence of non-recurring special events. This paper presents the methodology developed as part of this research.

Literature Review

Developing a standard approach for designing pedestrian traffic monitoring programs is progressing rapidly. Research performed in the area of pedestrian traffic monitoring has primarily focused on automated sensor performance, how long to conduct pedestrian counts, and characterizing traffic variation on multi-use paths. Recently, the National Cooperative Highway Research Program (NCHRP), Federal Highway Administration (FHWA), and Transportation Association of Canada (TAC) each published guidance on developing system-wide pedestrian traffic monitoring programs. These documents have combined existing literature to summarize and propose methods for jurisdictions to begin collecting pedestrian traffic data. The three documents, NCHRP Report 797 (*1*), the FHWA *Traffic Monitoring Guide* (*2*), and *Traffic Monitoring Practices Guide for Canadian Provinces and Municipalities* (*3*) were used to develop an initial approach for selecting continuous count locations in this research. Previous research on non-motorized site selection methods (*4*) was used to inform the specific placement of continuous count equipment in Winnipeg once TPGs were developed. Finally, quality control methods adopted for this study align with recommendations made by the FHWA's *Exploring Pedestrian Counting Procedures* (*5*). These include developing counter calibration factors, standards for acceptable amounts of missing data, and automated validation criteria such as comparing traffic records with historical counts and highlighting repeating values to be analyzed manually by a trained professional.

The purpose of a traffic monitoring program is to produce system-wide traffic statistics (*3*). Developing such a program begins with defining TPGs. For pedestrian traffic, the "system" is often limited to specific regions (e.g., multi-use path networks, commercial zones). TPGs are determined by sampling various locations in the network using short-duration counts. Similarities emerging from hourly and daily traffic patterns can be used to categorize sites into similarly behaving groups. Beyond this clustering approach, local knowledge and land use characteristics should be used to contribute an additional degree of pragmatism to the statistically determined categories (*3*).

Once TPGs are defined, the following steps include determining the number and location of continuous count sites, assigning these sites to appropriate TPGs, and creating a coverage count program with specific locations, frequencies, and durations (*3*). The ability to leverage short-duration counts to annual statistics is fundamentally dependent on assigning count sites to the appropriate TPG (*2*).

Selecting continuous count sites for active transportation traffic monitoring has been done in many jurisdictions in the past based on established TPGs. Miranda-Moreno et al. (*3*) developed two metrics used to assign bicycle counts to TPGs. These are the weekend-to-weekday index (WWI) and morning-to-midday index (AMI). The WWI quantifies the ratio of traffic on weekends compared with weekdays, whereas the AMI compares the traffic during the midday hours to morning hours. Together, these metrics are used to categorize sites as exhibiting recreational or utilitarian activity. These metrics are outlined below in equations 1 and 2.

$$AMI = \frac{V_{am}}{V_{mid}} \tag{1}$$

where V_{am} = mean morning peak (7:00 to 9:00 a.m.) traffic volume, and V_{mid} = mean midday peak (11:00 a.m. to 1:00 p.m.) traffic volume.

$$WWI = \frac{V_{we}}{V_{wd}} \tag{2}$$

where V_{wd} = mean daily weekday traffic, and V_{we} = mean daily weekend traffic.

For example, sites with a low WWI and high AMI indicate utilitarian use, as greater traffic occurs on weekdays and in the morning peak period. Conversely, recreational sites may be characterized by high activity during midday and weekends periods (*6*). Using this approach, locations within the network can be described and grouped by directly sampling specific time intervals. These metrics are important as they have become widely adopted for characterizing pedestrian and cyclist activity, particularly on multi-use path networks (*7*).

A gap observed in current research is determining the influence that special events have on pedestrian activity. Often falling outside the scope of research, it is common to categorize special events as a non-recurring variation, which is then omitted from coverage count data (*8*). These non-recurring events are typically omitted from coverage counts because traditional aggregated adjustment methods represent average conditions based on weekly and monthly averages. Although this approach is appropriate when events are infrequent, it is currently unclear how frequently non-recurring events must occur before they are considered by a traffic monitoring program. Day-of-year scaling factors developed by Hankey, Lindsey, and Marshall (*9*) can be used to overcome the limitations of traditional adjustment methods. Using this approach, daily coverage count data is referenced to the TPG's proportion of annual average daily traffic volumes during the count period. This approach develops a factor for each day of the year based on the percentage of annual average daily pedestrian traffic (AADPT) for all continuous count sites in the TPG.

A better understanding of the impact of special events in downtown areas is required to adequately plan for the safe and accommodating movement of pedestrians. Traffic conditions influenced by special events are important as they produce atypical demand on the transportation network. Negative impacts on safety and accessibility may result if this demand is not accounted for. Incorporating special event traffic also benefits traffic operations and planning tasks by informing design statistics such as the peak hour factor or 100th highest hour measures.

Study Area and Data Collection

The study area (Figure 1) is characterized by many office spaces, commercial buildings, and a large event arena. The arena's capacity is 15,000 and hosts events frequently

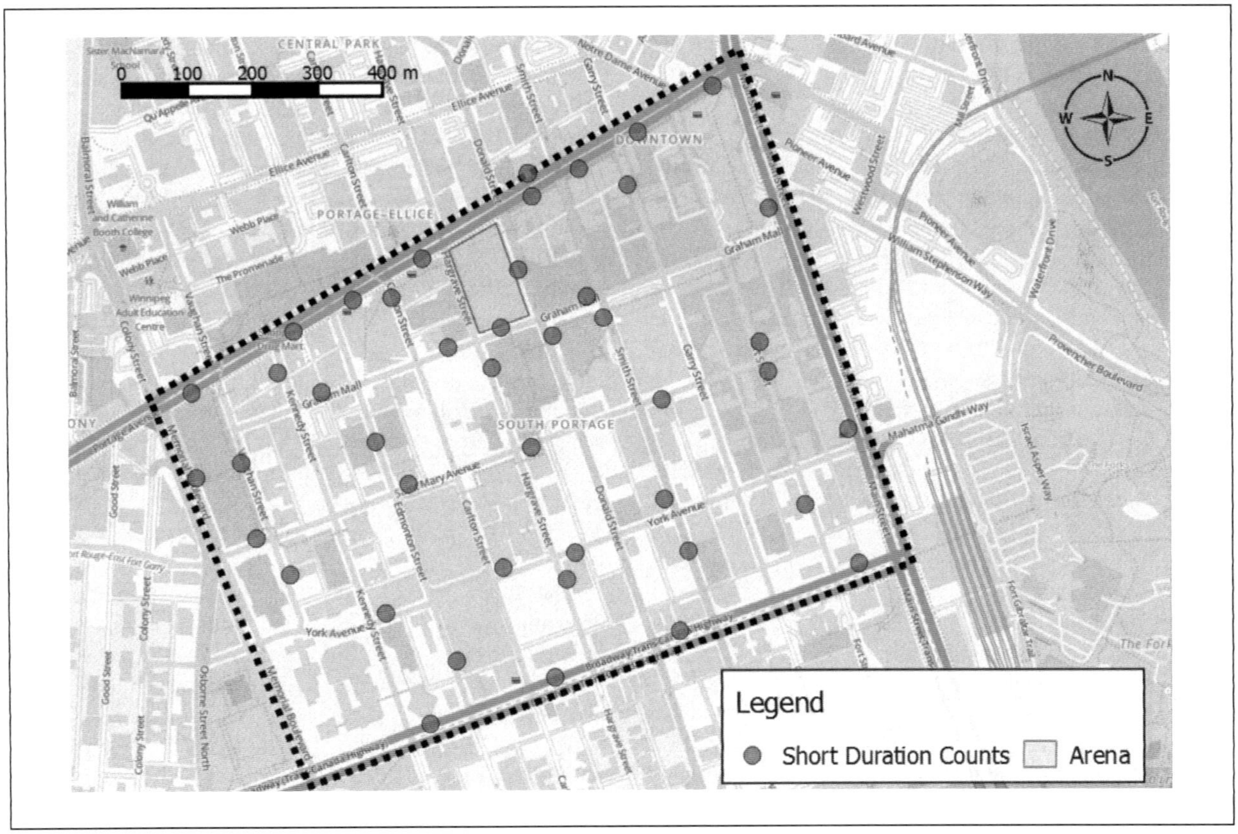

Figure 1. Study area and short-duration count locations.

Table 1. Automated Quality Control Flags

Flag name	Description
Consecutive zeros	If three consecutive records at a station are zero, the first occurrence of zero is flagged.
Consecutive count	If three consecutive records at a station are the same, the first occurrence of that value is flagged.
Average hourly	If the record is 70% greater or lesser than its corresponding annual average hourly weekday pedestrian traffic or annual average hourly weekend pedestrian traffic, the record is flagged.
Preceding count	If a record is 70% greater than the previous record, that hourly volume is flagged.

Source: Turner, Qu, and Lasley (*10*).

throughout the year (25 events over the 16-week study period). A short-duration count rotation was used to characterize pedestrian traffic in the study area. Eight infrared automated pedestrian counters (APCs) were used to conduct a (minimum) 7-day count at 45 locations. Count sites were evenly distributed across the study area subject to installation suitability. Specific locations were dependent on the availability of street-side mounting points and a non-reflective detection zone opposite the APC. The count rotation was conducted from July to November and resulted in 635 24-h pedestrian counts. Data from this count rotation included 45 unique count locations that collected traffic volumes over 89 days resulting in 15,240 h of recorded pedestrian volumes. This data set was used to analyze recurring and non-recurring trends in the study area.

Potential sources of error were identified and addressed by quality control measures. These sources were identified and addressed in post-processing. They included large over counts due to loitering pedestrians, undercounts due to sensor blockages, missing data due to sensor failure, or low counts due to construction or sidewalk closure. Automated flagging methods were used to highlight atypical values in the data set to minimize the effects of misrepresentative data. The purpose of the flagging procedure was to reduce the amount of manual effort required to review the raw data set. If a value in the dataset was flagged, it was removed if warranted by a manual inspection.

Standard thresholds were used to target approximately 5% of the most atypical pedestrian volumes in the dataset. This process was modeled after procedures used by Turner, Qu, and Lasley (*10*) to ease the manual data review burden for the data quality analyst. Due to varying pedestrian activity characteristics at each location, these thresholds occasionally were adjusted to flag manageable amounts of data for review. The quality control codes adopted from Turner, Qu, and Lasley (*10*) are listed in Table 1.

Using a procedure similar to Aultman-Hall, Lane, and Lambert (*11*), days with total precipitation over 5 mm were removed as wet weather has been observed to reduce pedestrian traffic. Each short-duration count was required to include at least each day of the week without precipitation. If precipitation occurred during the short-duration count, the APC was left installed until each day of the week was represented within the count period. The data resulting from this procedure was first used to describe each site's average weekday traffic (AWDT) and hourly distribution of pedestrian traffic. Figure 2 illustrates the AWDT at each site which ranged from 380 to 3,380 pedestrians per day. Figure 3 shows the average hourly weekday pedestrian traffic distribution at each site.

Quantification of Recurring Variation

Current pedestrian traffic monitoring practices and application of the WWI were insufficient for this research due to the frequency of special events in the study area. Using current methods, event days are omitted from short-duration count data to report typical volumes. If this approach were used, the frequency of special events at the arena would require many days of traffic to be omitted from the analysis of this research. Since it was observed that all special events hosted at the arena occurred after 7:00 p.m., the WWI was truncated to only consider traffic that occurs between 6:00 a.m. and 6:00 p.m. This allowed each day of data to be used as the influence of special events was removed from the WWI calculation. Volumes from Tuesdays through Thursdays were used to calculate the AWDT, whereas volumes from Saturdays and Sundays were used to calculate average weekend traffic. Excluding Mondays and Fridays was done to remove the influence of holidays when calculating AWDT (*12*).

The WWI and AMI were calculated at each site to determine whether all count locations behave similarly on a recurring basis. If this was found to be the case, further categorization of the study area could be made based on non-recurring variation.

The WWI compares average weekend traffic with AWDT. A WWI value greater than 1.0 is typical of recreational use, while a WWI less than 1.0 is typical of utilitarian use. The AMI analyzes peak periods within a

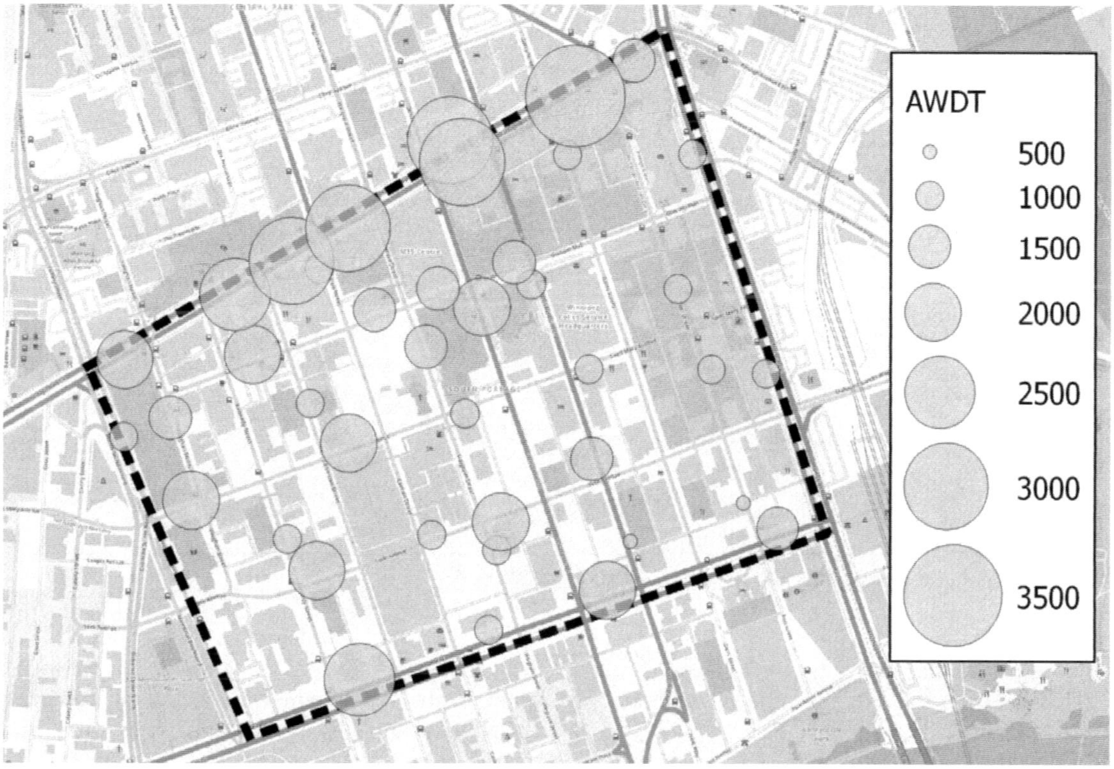

Figure 2. Average weekday traffic.

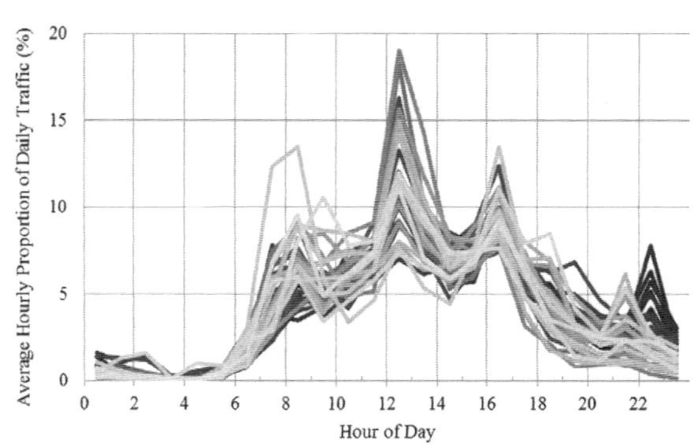

Figure 3. Average hourly proportion of daily traffic on weekdays.

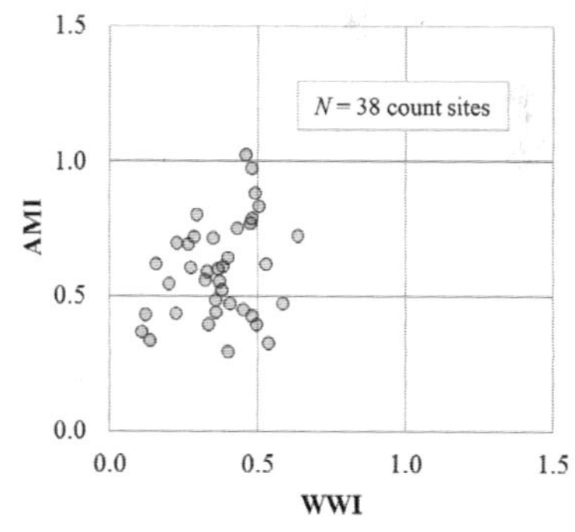

Figure 4. AMI vs. WWI of short-duration counts.

typical day, comparing the morning commute peak with the midday peak. An AMI greater than 1.0 is typical of utilitarian use, while values less than 1.0 are typical of recreational use. The WWI and AMI were calculated for each short-duration count. Figure 4 depicts the WWI–AMI coordinate for each count site. An AMI–WWI coordinate was not available at seven of the 45 sites due to data removed in the quality control process.

Given that the groups developed by Miranda-Moreno et al. (*6*) based bicyclist traffic data on multi-use paths

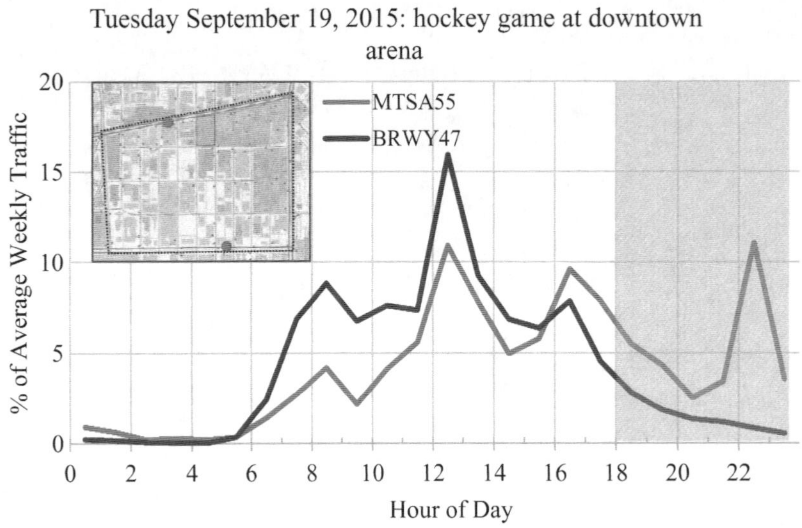

Figure 5. Event day: influenced site (MTSA55) vs. non-influenced site (BRWY47).

using all hours of the day and days of the week, it was expected that the truncated analysis of urban pedestrian traffic would yield different TPGs.

Figure 4 shows that the AMI value at each short-duration count site but one is below 1.0 and that all WWI values are below 0.75. According to the site classification criteria proposed by Miranda-Moreno et al. (*6*), each site would be characterized as mixed-recreational based on AMI values (i.e., more pedestrian traffic in the morning compared with midday) but characterized as utilitarian based on WWI values (i.e., more pedestrian traffic on weekdays compared with weekends). This research classifies all sites as exhibiting an "urban-utilitarian" pedestrian traffic pattern on a recurring basis where the AMI is less than 1.0 and the WWI is less than 0.75. This TPG is most likely not specific to Winnipeg's downtown. These results are to be expected wherever a large noon time peak is observed amidst traditional utilitarian traffic patterns, as described by Miranda-Moreno et al. (*6*). These conditions are commonly found in urban central business districts where commuter pedestrian activity is observed in close proximity to restaurants, green spaces, or other noon hour attractions.

Quantification of Non-Recurring Variation

An investigation was conducted to test whether sites behave differently during evening hours due to special events. Special events occurring at the downtown arena were identified using the arena's publicly provided event schedule. During the study period, on days used for analysis (Tuesdays to Thursdays), all events started at or after 7:00 p.m. An example of the difference between sites that were highly and minimally affected by special events hosted at the arena is shown in Figure 5. In this example, short-duration count site "MTSA55" is observed to be highly influenced by an event, whereas short-duration count site "BRWY47" is not. If sites were grouped only considering the AMI and WWI, this distinction would not be as evident as compared to a metric which considers evening traffic explicitly.

A metric, the evening proportion ratio (EPR), was developed by the authors of this research to categorize sites based on event influence. This metric considers hours of day from 6:00 p.m. to 11:59 p.m. This window is used to compare evening traffic between event days and non-event days while including the hour preceding the earliest event in the study period (6:00 p.m. to 7:00 p.m.). Including only an hour before event time was found to adequately encompass the increase of pedestrian traffic due to an event. Other municipalities may find that event traffic increases two or more hours before an event due to a higher concentration of restaurants or other pre-event activities. These municipalities may want to increase the EPR time period.

Determining EPR to quantify event influence requires several steps. First, the total evening pedestrian volume (6:00 p.m. to 11:59 p.m.) is divided by the total daily volume. This produces the proportion of that day's volume that occurs during the evening period. This proportion is expressed as a percentage of daily pedestrian traffic and referred to as the evening proportion index (EPI). By dividing the average EPI of event days by the average EPI of non-event days, the influence of special events at a count

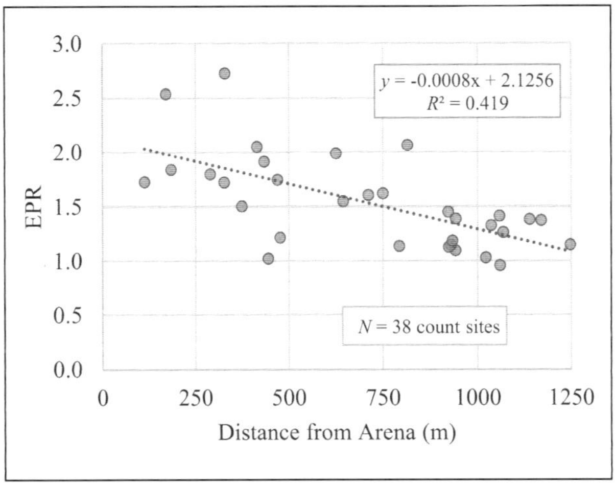

Figure 6. Sidewalk distance vs. EPR.

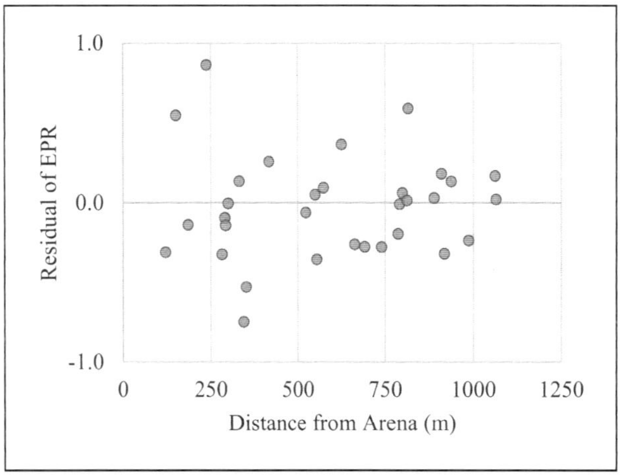

Figure 7. Residual of EPR vs. distance.

site can be evaluated. This value is referred to as the EPR. The EPI and EPR are described in equations 3 and 4.

$$\overline{EPI} = \frac{1}{n}\sum_{i=1}^{n}\sum_{j=18}^{24} \%Daily\ Volume_{ij} \quad (3)$$

where

n = number of days in short duration count,
i = day in the counter duration,
j = hour of day.

$$EPR = \frac{\overline{EPI}_{event}}{\overline{EPI}_{non-event}} \quad (4)$$

where

EPI_{event} = EPI calculated on event days,
$EPI_{non-event}$ = *EPI calculated on non event days*,
$\% \ Daily\ Volume_{ij}$ = j^{th} hourly proportion of daily traffic on the ith day of the count.

The resulting EPR value for each short-duration count is an indication of the degree to which that site is influenced by special events hosted at the arena. If the value of the EPR is equal to 1.0, this indicates that a site is not influenced by events hosted at the downtown arena. As the EPR value increases above 1.0, this indicates that a greater level of evening pedestrian traffic occurs on event days compared with non-event days. An EPR value < 1.0 may indicate that the site is influenced by a different venue that holds events on different days.

The EPR makes it possible to assess the degree that each short-duration count site was influenced by events occurring at the downtown arena. It was hypothesized that there would be a negative correlation between distance from the arena and the EPR value. Once this correlation was assessed and the influence of events was mapped, the study area could be separated into differing influence zones. Based on the emerging trends within the study area, the study area was categorized into different TPGs, each receiving an equal number of continuous count sites.

Results

Each count site was initially classified as "urban utilitarian" based on recurring traffic variation. The analysis determined each site's response to special events using the EPR. Figure 6 illustrates the relationship between EPR and each count site's direct distance from the arena. From this regression analysis, the study area could be categorized into different groups of varying levels of influence to define TPGs based on non-recurring variation.

A negative relationship between distance and event influence is shown in Figure 6 with moderate correlation (R^2 = 0.419). Given that the points of the residual plot (Figure 7) are randomly dispersed about the horizontal axis, it can be concluded this relationship is approximately linear.

To visualize the relationship between event influence and distance to the downtown arena spatially, Figure 8 shows a map of the EPR calculated from each short-duration count. Orange symbols indicate that the EPR value at that location exceeds 1.5, which designate sites as being highly influenced by events at the arena. Values of the EPR for this analysis fell between the values of 0.91 and 3.48. This range indicates a wide variation in the degree to which locations are influenced by special events. The most influenced site on average has 3.48 times greater traffic on event days, however, one site actually experienced less traffic on average on event days by a factor of 0.91.

Figure 8 shows a concentration of highly influenced sites near the arena and along the north border of the study area. Furthermore, as distance from the arena increases, the density of highly influenced sites decreases.

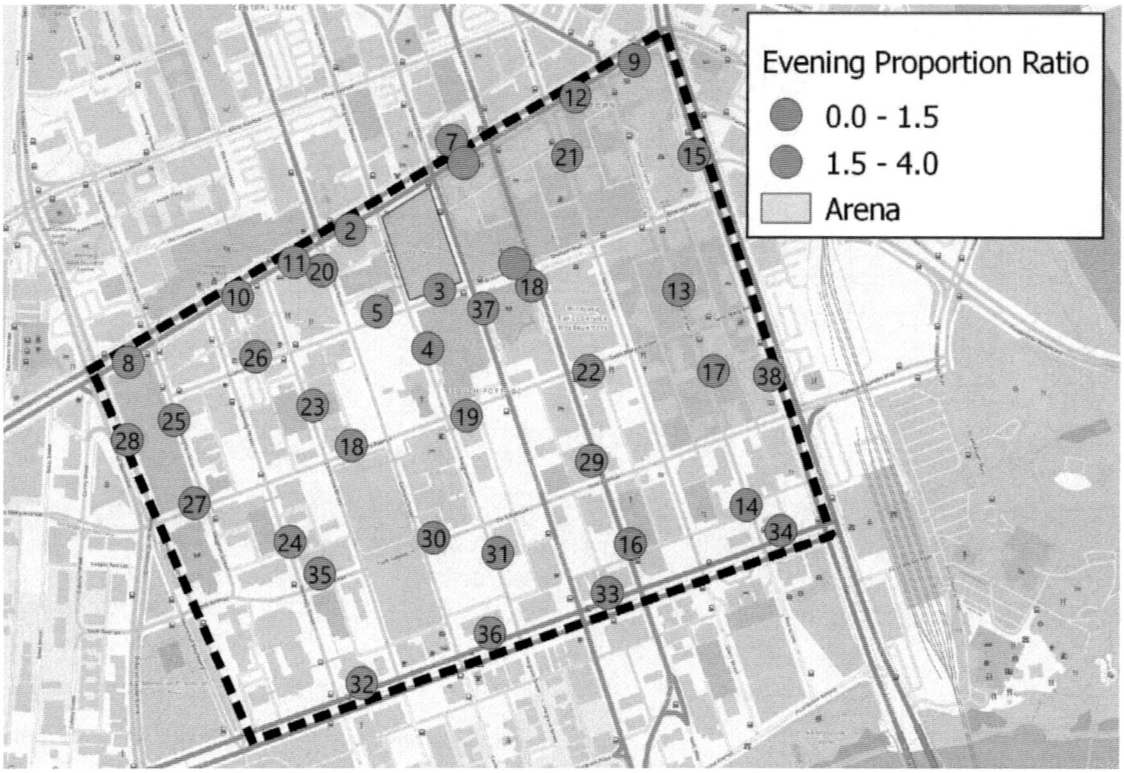

Figure 8. Evening proportion ratio for short-duration counts.

Using each site's WWI, AMI, and EPR values, the study area was categorized into two separate groups for the purpose of selecting continuous count sites. The AMI, WWI, and EPR values for each count site are tabulated in Table 2.

Discussion

The purpose of this research was to investigate the most suitable continuous count locations within the study area. These continuous counts would serve as reference for AADPT adjustment for coverage counts performed throughout the study area by the city of Winnipeg. The FHWA *Traffic Monitoring Guide* (*2*) states that at least three continuous counts are required per TPG for pedestrian traffic to accurately determine AADPT. Given the distribution of high and moderate event influenced sites, two TPGs were defined spatially called the "urban utilitarian – event" and "urban utilitarian" TPGs respectively. Each TPG was assigned four continuous counters.

Figure 9 illustrates two proposed zones within the study area; one considered to be highly influenced by special events hosted at the arena and one which is moderately influenced. Since installation, the eight continuous counters have collected over a year of pedestrian volumes. The city of Winnipeg's development of a pedestrian monitoring program that adjusts coverage counts is ongoing. Once a short-duration count schedule is developed, this research recommends all counts within the "highly influenced" zone be expanded by continuous counts of the "urban utilitarian – event" TPG, and all other coverage counts be expanded by the continuous sites of the "urban utilitarian" TPG.

A few limitation of this research are inherent to the data collection process. First, due to resource and funding availability, the study was limited to the use of eight APCs. In order to sample the number of sites in this study, the data collection method required these counters to be rotated to different locations throughout the study period., The results would be strengthened if counts at each site were conducted for a full year. Future research is needed to quantify the accuracy of average traffic statistic estimates from short-duration count data using this methodology of developing pedestrian TPGs.

To reiterate, the primary goal of this study was not to develop precise descriptive statistics for each study site, but to develop a method for using a short-duration count rotation to select continuous count sites based on observed TPGs. This study completes a preliminary characterization of pedestrian traffic patterns in a

Table 2. Average Weekday Traffic Metrics by Count Site

Station No.	1	2	3	4	5	6	7	8	9
WWI	0.45	0.53	0.64	0.34	0.48	0.33	0.46	0.49	0.41
AMI	0.45	0.33	0.73	0.39	0.97	0.59	0.94	0.40	0.47
EPR	1.73	1.84	2.73	1.91	1.74	1.50	2.54	2.06	1.99
Station No.	10	11	12	13	14	15	16	17	18
WWI	0.38	0.36	0.43	0.14	0.23	0.28	0.12	0.37	0.32
AMI	0.61	0.44	0.75	0.34	0.69	0.61	0.43	0.55	0.56
EPR	2.86	3.48	2.05	1.09	1.14	1.13	1.03	1.21	2.18
Station No.	19	20	21	22	23	24	25	26	27
AMI	0.58	0.48	0.35	0.53	0.27	0.11	0.29	0.40	0.40
AMI	0.47	0.77	0.72	0.62	0.69	0.37	0.80	0.30	0.64
EPR	1.55	1.72	1.02	1.60	1.62	2.08	1.39	1.13	1.37
Station No.	28	29	30	31	32	33	34	35	36
AMI	0.48	0.50	0.22	0.38	0.16	0.29	0.37	0.36	0.20
AMI	0.43	0.83	0.44	0.52	0.62	0.72	0.60	0.49	0.55
EPR	0.96	1.18	1.26	1.45	0.91	1.13	1.26	0.98	2.10
Station No.	37	38							
AMI	0.49	0.48							
AMI	0.88	0.79							
EPR	1.22	2.46							

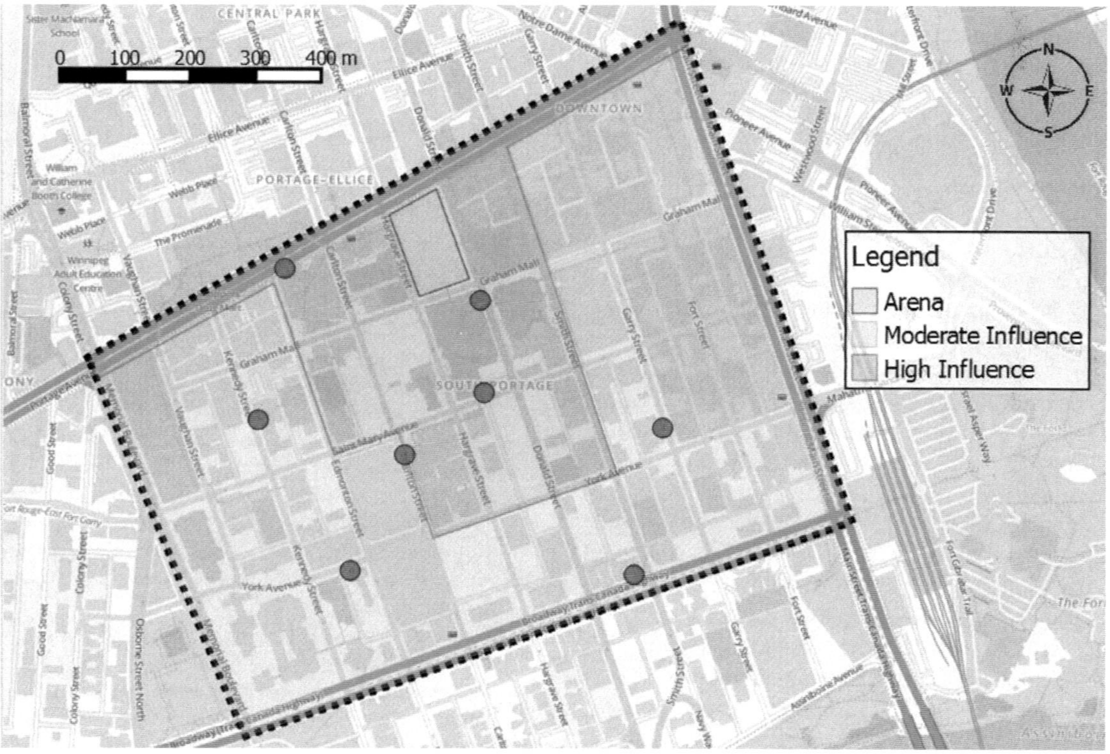

Figure 9. Event influence zones with continuous count locations.

relatively homogeneous urban environment to guide the selection of continuous count sites. As such, the methodology was optimized to maximize network coverage while recognizing the potential loss of accuracy for group statistics (e.g., AMI, WWI, and EPR). Future research should also confirm and evaluate the accuracy of the AMI, WWI, and EPR variables used to define the pedestrian TPGs.

Conclusion

This paper outlines a method for selecting continuous count sites in highly urbanized environments by considering the influence of special events. The research leading to this paper builds upon existing knowledge by addressing frequently occurring large-scale special events. An understanding of recurring patterns was developed by using short-duration counts, and count sites were characterized based on their recurring variation. Once the similarity between sites was established, the sites were further described by their traffic variation due to events hosted at the downtown arena using the EPR.

It was determined that daytime traffic at each site can be classified within the "urban-utilitarian" TPG, which exhibits a WWI value less than 0.75 and an AMI value less than 1.0. The study area was then subdivided into two areas: one that was highly influenced by events at the downtown arena, and one that was only moderately influenced. Within each of these two areas, four APCs were installed to collect continuous data for a full year. These eight continuous count sites will serve as the basis for an ongoing pedestrian monitoring program.

As downtown areas typically experience the highest pedestrian traffic volumes in a city, these areas provide an opportunity for the greatest return on traffic monitoring and infrastructure investments. This research presents a systematic approach to accurately measure pedestrian traffic in these areas that are often first to be targeted by pedestrian accommodation improvement programs. Incorporating the effect of special events explicitly in a downtown traffic monitoring program is critical to providing planners and engineers with reliable data for design that accommodates all modes. For example, recognizing that peak pedestrian volumes may not fall within traditional peak periods (7:00 to 9:00 a.m., 3:00 to 5:00 p.m.) or during the noon peak period (11:00 a.m. to 1:00 p.m.) has significant impacts for how facility demand and level of service are understood. Furthermore, pedestrian facilities may be under designed if crossing exposure for safety analyses are evaluated during traditional peak periods, rather than when the largest pedestrian volumes actually occur. In the Winnipeg context, the data being collected from this pedestrian traffic monitoring program have been used to measure changes in pedestrian traffic after adopting a city-wide pedestrian and cyclist accommodation strategy.

This research builds a case for considering special events in planning pedestrian facilities. Although Winnipeg's downtown pedestrian traffic is significantly affected by a single large event center in the evenings, an opportunity for future research is to consider the impact of multiple proximate event centers, or events that are not limited to evening hours.

Author Contributions

The authors confirm contribution to the paper as follows: study conception and design: CO, RP, JM; data collection: CO, RP; analysis and interpretation of results: CO, RP; draft manuscript preparation: CO. All authors reviewed the results and approved the final version of the manuscript.

References

1. Ryus, P., E. Ferguson, K. M. Laustsen, R. J. Schneider, F. R. Proulx, T. Hull, and L. Miranda-Moreno. *NCHRP 797: Guidebook on Pedestrian and Bicycle Volume Data Collection.* Transportation Research Board of the National Academies, Washington, D.C., 2014.
2. Federal Highway Administration. Traffic Monitoring Guide. U.S. Department of Transportation. Washington D.C., 2013.
3. Regehr, J. D., R. Poapst, G. Rempel, J. Montufar, and M. Hallenbeck. *Traffic Monitoring Practices Guide for Canadian Provinces and Municipalities.* Transportation Association of Canada, 2016.
4. Jackson, K. N., E. Stolz, and C. Cunningham. Non-Motorized Site Selection Methods for Continuous and Short-Duration Volume Counting. *Transportation Research Record: Journal of the Transportation Research Board,* 2015. 2527: 49–57.
5. Nordback, K., S. Kothuri, T. Petritch, P. McLeod, E. Rose, and H. Twaddell. *Exploring Pedestrian Counting Procedures.* Federal Highway Administration, U.S. Department of Transportation. 2016.
6. Miranda-Moreno, T. Nosal, R. Schneider, and F. Proulx. Classification of Bicycle Traffic Patterns in Five North American Cities. *Transportation Research Record: Journal of the Transportation Research Board,* 2013. 2339: 49–57.
7. Beitel, D., and L. Miranda-Moreno. Methods for Improving and Automating the Estimation of Average Annual Daily Bicyclists. Presented at 95th Annual Meeting of the Transportation Research Board, Washington, D.C., 2015.
8. Nordback, K., W. E. Marshall, and B. N. Janson. *Development of Estimation Methodology for Bicycle and Pedestrian Volumes Based on Existing Counts.* Publication CDOT-2013-18. Colorado Department of Transportation, 2013.
9. Hankey, S., G. Lindsey, and J. Marshall. Day-of-Year Scaling Factors and Design Considerations for Non-Motorized Traffic Monitoring Programs. *Transportation Research Record: Journal of the Transportation Research Board,* 2014. 2468: 64–73.
10. Turner, S., T. (T.) Qu, and P. Lasley. *Strategic Plan for Non-Motorized Traffic Monitoring in Colorado.* Texas Transportation Institute, Texas A&M University System, 2012.
11. Aultman-Hall, L., Damon L., and R. R. Lambert. Assessing Impact of Weather and Season on Pedestrian Traffic Volumes. *Transportation Research Record: Journal of the Transportation Research Board,* 2009. 2140: 35–43.
12. Poapst, R. Characterizing Pedestrian Traffic by Hour-of-Day Periodicities in Commercial Zones. MSc thesis. University of Manitoba, Winnipeg, 2015.

The Standing Committee on Highway Traffic Monitoring (ABJ35) peer-reviewed this paper (18-01261).

Article

TRR
JOURNAL OF THE TRANSPORTATION RESEARCH BOARD

Transportation Research Record
2018, Vol. 2672(43) 75–86

DOI: 10.1177/0361198118783123
journals.sagepub.com/home/trr
SAGE

Automated Validation and Interpolation of Long-Duration Bicycle Counting Data

David Beitel[1], Spencer McNee[1], Fraser McLaughlin[2], and Luis F. Miranda-Moreno[1]

Abstract

Bicycle flow data is crucial for transportation agencies to evaluate and improve cycling infrastructure. Average annual daily bicyclists (AADB) is commonly used in research and practice as a metric for cycling studies such as ridership analysis, infrastructure planning, and injury risk. AADB is estimated by averaging the daily cyclist totals measured throughout the year using a long-term automated bicycle counter, or by using long-term bicycle counting data to extrapolate data from a short-term counting site. Extrapolation of a short-term bicycle counting site requires an accurate and complete set of daily factors from a group of references: long-term bicycle counters. In practice, validation of reference data is done manually, an exercise that is time-consuming but crucial as significant error can be introduced into AADB extrapolation if reference data are not validated. This paper proposes an automated method to validate long-term bicycle count data and interpolate anomalous portions of data. As part of this work, the methods are validated using a relatively large dataset of automated bicycle counts. For validation of our approach, data anomalies are created artificially in a way that removes data (first trial), or reduces counts to 25% or 40% of the measured bicycle counts (second and third trials), for 6 hours, 12 hours, and full days. Of the more than 100 generated anomalies, the validation process flagged approximately 90% in the first and second trials and 80% in the third trial. The average absolute relative error of the interpolated daily values was approximately 10% for all three trials.

Bicycle flow data is crucial for transportation agencies to evaluate and improve cycling infrastructure. Transportation agencies can use bicycle flow data at locations across a network to help prioritize intersections awaiting road treatments, to determine which intersections require bike signal phasing, and to estimate exposure in traditional crash-based safety analysis for cyclists (*1*). Measuring flow and estimating daily averages is considerably simpler for motorized vehicles as compared to bicycles. Firstly, cyclists are harder to count than vehicles using traditional automated counting systems; vehicles are big and typically occupy a predictable space making them easy to count using automated counting technologies. Secondly, vehicle flows tend to be large and daily volume is stable over time, therefore short-duration counts can be used to predict daily averages with a relatively high-level of accuracy. Bicycle flows, on the other hand, tend to be smaller and vary significantly as a function of time and weather. Thus, the estimation of daily averages from short-duration counts is a laborious process and prone to error.

Average annual daily bicyclists (AADB) is the standard metric used by researchers and practitioners to quantify cycling flow, referred to as exposure, volume, ridership, or activity on bicycle facilities, at intersections or on road sections (*2*, *3*). Automated bicycle counting systems, used for measuring bicycle ridership, fall into two categories: long-term counters that run continuously at one location for multiple years, and mobile counters that capture a short duration of cycling demand (typically from 1 day to a month, and ideally for longer than a week) (*3*). The data from long-term counters can be used to calculate the AADB at a location when installed for more than 1 year, simply by averaging the daily bicycle demand throughout the year or bicycle season. Estimating the AADB with a short-term count is possible but requires extrapolation methods using knowledge of daily cycling demand patterns from appropriate references: long-term counting sites (*3*, *4*). Direct measurement of AADB using a long-term counter can be highly accurate but expensive. On the other hand, estimating AADB with a short-term count and extrapolation techniques, although relatively inexpensive, will produce results of varying accuracy. The accuracy of an AADB estimation is known to vary with several characteristics of the short-term count, including the

[1]Department of Civil Engineering and Applied Mechanics, McGill University, Montréal, QC, Canada
[2]Eco-Counter Inc., Montréal, QC, Canada

Corresponding Author:
Address correspondence to David Beitel: david.beitel@mail.mcgill.ca

time of year the data were collected, the count duration, and the level of cycling demand (*3*, *5*), as well as the goodness of match between the short-term count and the reference(s) used in extrapolation (*6*).

The accuracy of AADB estimation from short-term counts is also highly susceptible to error in the reference data, particularly in cases in which the reference experiences significant undercounting during the same period that the short-term count is taken. Undercounting of the reference count data can be difficult to identify if the counts are significantly reduced but non-zero. Undercounting can have many causes including automatic counter malfunction, temporary loss of power, and construction on or near cycling facilities. The most common source of undercounting may be the presence of utility, moving, or other vehicles diverting bicycles around the counting site.

At present, long-term bicycle count validation is done manually in practice, if at all, by researchers and practitioners. It is an arduous task involving many steps; however, it is crucial as long-term bicycle counting sites form the backbone of any city- or region-wide counting program. They provide rich data sets that can be used to evaluate bicycle ridership growth, in addition to being used as references to estimate AADB from short-term count data. However, when extrapolation is done, any errors in the long-term count data will be magnified as error in the AADB estimate of the short-term counting site. This is particularly true of zero- or highly suppressed counts in the long-term counts (*6*). An automated process that runs through all the steps of bicycle count data validation of long-term reference sites, by flagging and removing anomalous data and then replacing it through interpolation, would be useful in comprehensive bicycle counting programs and would make the AADB estimation from short-term counts more feasible in practice. This study builds on previous work related to semi-automated data validation of permanent bicycle counts (*6*). It has been demonstrated that nearby reference counting sites can be grouped together if their temporal cycling patterns are similar (*4*, *6*, *7*). Groups of similar counting sites can subsequently be used to identify periods of anomalous count data and to correct data during those periods.

The study objectives are to develop and fully automate a methodology for data validation and data interpolation to be used on reference counting data; and to test the methodology on long-term bicycle count data with simulated zero-count or suppressed-count anomalies. The simulated anomalies are limited to reduced or zero counts; however, this methodology also applies to anomalies that increase counts. The data used in this study come from 16 long-term bicycle counting sites in Arlington, VA. The proposed methodology is the result of collaboration with Eco-Counter, a manufacturer of automatic bicycle and pedestrian counting equipment. The methodology has been incorporated into software that will be accessible to jurisdictions with Eco-Counter equipment, but could be applied to continuous daily bicycle count data collected from any bicycle counting system.

Literature Review

A complete bicycle counting program, administered by a city, regional or state/provincial transportation agency, requires both long-term reference counting sites as well as short-term counting sites. A growing body of work has demonstrated several bicycle demand extrapolation techniques, with varying accuracy, for short-term counting sites (*3*, *6–15*). Nordback et al. demonstrated that average annual daily traffic extrapolation techniques for motor vehicle counting can be borrowed to estimate AADB (*3*). These techniques use day-of-week and month-of-year factors developed using continuous count data, taken over an entire year, from a group of counters located at similar cycling facilities. The authors explored how the error in AADB estimation is related to the duration of the short-term count, the month in which the short-term count was taken and the average bicycle demand. Nosal, Miranda-Moreno, and Krstulic examined four methods for AADB extrapolation including the disaggregated factor method, also known as the day-of-year factor method, which produced the best results (*8*). In this method, an expansion factor is computed for each day of the year using the raw daily counts and the annual, or seasonal daily average of a long-term reference counting site. This set of factors would typically be produced using a long-term counting site near the short-term counting site that requires extrapolation. The idea behind this method is that the proximity of the two counting locations ensures that weather affects cycling demand at both locations to a similar extent. The average absolute relative error (AARE) was compared for different extrapolation methods for short-term counts ranging from 1 day to 1 month's duration.

Esawey studied AADB estimation error from short-duration counts using automated bicycle counters (*11–14*). In one study, daily and monthly adjustment factors were used to estimate AADB; estimation errors were then compared across different days of the week and across different months of the year (*12*). In another study, Esawey compared the estimation accuracy of methods used to estimate AADB including both traditional factor methods and day-of-year factor methods (*14*). Fournier proposed a sinusoidal model with a calibration factor to adjust for impact of seasonal demand change (*16*). The model can estimate monthly average daily bicycle counts and AADB using as little as two short-term counts, at high- and low-points in the cycling season. Hankey, Lindsey, and Marshall compared the day-of-year factor method to the more traditional day-of-week and month-of-year factor methods using data from off-street trail locations and found a significant reduction in AADB extrapolation error using the day-of-year factor method (*9*). A couple of studies have explored how AADB extrapolation error could be further reduced by correcting for specific sources of error. Beitel and Miranda-Moreno explored two approaches for extrapolating AADB that improved on previously reported work on the day-of-year factor method, and achieved more consistently accurate estimates than previously reported methods (*6*, *8*, *9*). One method utilized a filtering technique to remove AADB estimate outliers (Beitel and Miranda-Moreno, 2016). Figliozzi et al. proposed a methodology to reduce AADB extrapolation error by applying a correction function that used a regression model to account for weather and non-school days (*10*).

Several studies have demonstrated extrapolation techniques to estimate AADB from short-term counts (*3*, *6*, *9*, *11–16*). Each study assessed the accuracy of the extrapolation method developed using AARE (referred to in other studies as mean average percent error or MAPE) and some have compared errors for different methods developed in the literature (*6*, *8*, *9*, *12*, *14*). Furthermore, several studies have explored the relationship between AADB estimation error and short-term count duration, bicycle demand, time-of-year of the short-term count, and other factors (*3*, *5*, *9*, *14*). Some recent contributions have addressed important research gaps related to site selection and implementation of bicycle counting programs (*17–19*) and methodologies to assist in quality assurance and quality control (*19–21*). The *Traffic Monitoring Guide*, produced by the U.S. Federal Highway Administration, provides guidance on non-motorized counting programs, describes challenges in collecting non-motorized data, and includes case studies that define manual quality-control techniques that can be applied to bicycle count data (*19*). The national bicycle and pedestrian count archive, called Bike-Ped Portal, completed on behalf of the National Institute for Transportation and Communities, allows for nationwide data sharing and offers some built-in features including data visualization and zero-data detection (*22*). These contributions have helped make AADB extrapolation from short-term counts more feasible. However, a significant research gap remains that limits the applicability of AADB estimation from short-term counts in practice. No studies have developed a robust, automated technique for the validation of long-term reference count data.

Validating and correcting reference bicycle count data is crucial in achieving a high accuracy in AADB extrapolation, regardless of the extrapolation method subsequently used. Undercounting of the reference data can have a large impact on the accuracy of the daily AADB extrapolation of a short-term counting location because the daily bicyclist factors of the reference data appear in the denominator of the estimate (*6*). Therefore, a partial restriction of a bicycle facility, at a reference counting location, that reduces the bicycle counts by 75% during the period of the short-term count would produce an AADB estimation error at the short-term count location of 300%. In estimating AADB, Nosal manually removed all zero and missing data, resulting in a loss of 4% of data (*8*). Currently, the only available method for identifying undercounting and other anomalies is to manually analyze continuous data by hourly intervals. This is typically done by viewing data graphically; all anomalies that cannot be explained by variation in local weather are then flagged or removed (*4*, *6*). This method, though highly valuable, is arduous and time-consuming. It can also be challenging to spot data that is highly reduced yet non-zero due to a partial blockage of the bike facility. An automated method would be significantly less time-intensive and more complete than the manual methods that are used in practice. The automated validation method proposed in this study is an extension of a semi-automated validation method proposed by Beitel (*6*). In the case that data from multiple reference counters are available for an entire year (or cycling season) in a single city or region, then data validation and anomaly identification can be achieved by validating the daily factors of each counting site against the daily factors of other similar counting sites. An automated process that validates reference bicycle count data would provide practitioners with a tool that can help in the estimation of bicycle flows at many locations across city- or region-wide bicycle networks.

Methodology

Anomalous data are easily identified as periods when a counting location has a variation about its daily factor that is not mirrored at other, similar counting locations. The proposed process identifies similar counting sites for each reference site, detects days of the cycling season when similar sites have differing daily factors, and, if necessary, removes and interpolates the anomalous daily factor. The process consists of the following six steps:

(1) *Calculate the daily factors for each long-term counting location.* Each daily factor is calculated by taking the quotient of the bicycle count on that day and the annual, or seasonal, daily average. Note that the method for determining the daily factors was proposed by Nosal as part of the day-of-year factor method for estimating AADB and is used in this study in a different context, for validating reference data (*8*).

(2) *Calculate the correlation between all possible pairs of reference sites.* For each site i, calculate the correlation between the vector of daily factors and the vector of daily factors for each other site j, according to Equation 1 and simplified to Equation 2.

$$Correl\left(DF_i, DF_j\right) = \frac{\sum_d \left(df_{i,d} - \overline{df_{i,d}}\right)\left(df_{j,d} - \overline{df_{j,d}}\right)}{\sqrt{\sum_d \left(df_{i,d} - \overline{df_{i,d}}\right)^2 \sum_d \left(df_{j,d} - \overline{df_{j,d}}\right)^2}} \quad (1)$$

where

$Correl(DF_i, DF_j)$ varies between 0 and 1, being 1 when factors are perfectly correlated,

DF_i, DF_j represent the vector of daily factors for the reference counting sites i and j respectively,

$df_{i,d}, df_{j,d}$ represent the daily factor on day d for the reference counting sites i and j respectively.

As the average daily factor is equal to 1 for any site, the formula simplifies to,

$$Correl\left(DF_i, DF_j\right) = \frac{\sum_d \left(df_{i,d} - 1\right)\left(df_{j,d} - 1\right)}{\sqrt{\sum_d \left(df_{i,d} - 1\right)^2 \sum_d \left(df_{j,d} - 1\right)^2}} \quad (2)$$

where

(3) *Select similar counting sites.* For each reference site i, select up to two counting sites j with the highest correlation of daily factors calculated in Step 2. For a similar counting site to be chosen, the correlation between the daily factors must be greater than a threshold, corr_min (in this study, corr_min = 0.75).

(4) *Compute the ratio of daily factors.* For each day of the counting season, calculate the ratio, df_i / df_j, for the two similar counting sites j selected in Step 3.

(5) *Set threshold and flag days.* Set a value for e (in this study, $e = 2$), then, for both sets of ratios computed in Step 4, flag all daily ratios that are outside of the interval $[1/e, e]$.

(6) *Remove and interpolate.* All reference daily factors on days that are flagged in both sets of ratios from Step 5 are removed and interpolated. The interpolated daily factor is the average of the daily factors of the two similar counting sites.

The automatic data validation and interpolation program was tested on Arlington data. Several values of each software parameter were tested first to determine optimal values for use in the study. The corr_min value, used in Step 3 of this procedure, is 0.75. Therefore, for a counting site to be validated using similar sites, the correlation of the daily factors of the two sites is required to be at least 0.75. If the correlation is lower, the site under scrutiny is considered excessively divergent for validation purposes. A corr_min value of 0.75 strikes a balance between being overly restrictive to sites with multiple days of anomalous data and overly permissive in allowing sites with extremely different temporal profiles to be used as references for one another. Note that as the number of days that a counting site has anomalous (zero, reduced, or inflated) data in a year increases, its correlation with other counting sites tends to decrease. This relationship places an upper limit on the number of days of missing data a counting site can have in a year. The threshold used in this study was 15 days, and is described in the Data section.

The e value, used in Step 5 of the procedure, is 2. This means that the ratio of daily factors between two counting sites that falls outside of the interval [0.5, 2] is flagged for removal and replacement by interpolation. Different values of e were tested prior to this study. Lower values of e (for example, $e = 1.5$) set a lower threshold for triggering the flag and the interpolate feature of the software. Lower values of e have the advantage of being more sensitive to anomalous data and can detect partial days of anomalous counts, however, they are prone to false positives, i.e., flagged non-anomalous daily factors from inherent variation of counts. On the other hand, higher values of e (for example $e = 2.5$) set too great a threshold and tend not to detect days with counts reduced by as much as 60%. The validation procedure was repeated for values of e ranging from 1.5 to 2.5 in 0.1 increments. The value of $e = 2$ was found to be optimal with the test data because false positives were eliminated whereas the filter preserved a strong sensitivity to anomalous data; full days of simulated anomalous data were flagged more than 19 times out of 20 (see Results and Discussion section).

Table 1. Example of Correlation Matrix Calculated in Step 2 and Used in Step 3 of Validation Process

Counting site	Site 1	Site 2	Site 3	Site 4	Site 5	Site 6
Site 1	na	0.828	0.916	0.914	0.937	0.969
Site 2	0.828	na	0.930	0.944	0.921	0.787
Site 3	0.916	0.930	na	0.979	0.944	0.872
Site 4	0.914	0.944	0.979	na	0.962	0.890
Site 5	0.937	0.921	0.944	0.962	na	0.918
Site 6	0.969	0.787	0.872	0.890	0.918	na

Note: na = not applicable.

Table 1 presents a matrix of correlation values calculated between all i–j counter pairs. This is an example that consists of six counting sites. Each row has two shaded boxes corresponding to the two most similar reference sites to the counting site representing the ith row, determined by having the highest correlation.

Error Measurement

The accuracy of a daily factor estimation is defined by the absolute relative error (ARE) as given in Equation 3. The flagged daily factors are removed and estimated through interpolation. The ARE of each daily factor interpolation is reported in the Results and Discussion section.

$$ARE = \left|Error_{i,y,j}\right| = \frac{\left|\widehat{df}_{i,j,y} - df_{i,j,y}\right|}{df_{i,j,y}} \quad (3)$$

where

$\left|Error_{i,y,j}\right|$ = the ARE for reference bicycle counting site i, based on the estimated and observed daily factors on day j of cycling season y,

$\widehat{df}_{i,y,j}$ = the estimated daily factor for reference site i on day j of cycling season y,

$df_{i,y,j}$ = the observed daily factor for reference site i on day j of cycling season y.

Data

The continuous bicycle count dataset used in this analysis was obtained from inductive loop bicycle counters manufactured by Eco-Counter and owned by Arlington County. Data from this equipment has been used in a wide range of studies and, when operating properly, ARE of these counters has been shown to be below 4% (*2*, *23–25*).

The data used to test the methods developed in this study came from 16 long-term bicycle counters in Arlington. The bicycle demand data in the analysis includes both weekday and weekend data for the 2015 and 2016 cycling seasons. The cycling season, which was defined in this study as all months where the cycling demand was at least 50% of the peak season demand, is from April 1 to November 30 in Arlington. Counting sites were only considered for the study if the AADB was greater than 250. Four counters in the Arlington dataset were disqualified from analysis for not satisfying that condition, bringing the number of eligible counters down to 16 from 20. The choice of cycling season months, with a threshold of 50% retention rate of the peak-demand month, and the minimum AADB condition were chosen to ensure that the daily bicycle counts at each counting site were large enough for validation purposes. As bicycle counts decrease, the proposed filtering method becomes less effective as the relative temporal variation in counts tends to rise rapidly. These two conditions represent limitations on when and where the filtering method can be applied.

Data Preparation

The automated validation and interpolation process has two stages. In the first stage, the data is scanned for extended periods of zero counts or missing data. The validation and interpolation program includes a parameter: the zero or missing data threshold. If a counting site has missing or zero data for more days than the threshold permits, then that counting site is removed from analysis. The threshold was set at 15 days in this study. This threshold was selected to agree with the corr_min value described in the Methodology section. If a counting site has more than 15 days of missing or zero data in a year, then the correlation of daily factors with other counting sites is not high enough for validation purposes. In the second stage, the ratio of daily factors for each counting site to the daily factors of the two closest references, as explained in Step 3 of the Methodology section, are taken to flag any days with anomalous bicycle counts. Each counting site in the analysis was validated and interpolated prior to testing the process with anomaly simulations. The first stage of the process removed two counters from analysis in 2015 and one in 2016. In the second stage seven and eight counter-days were flagged and interpolated in 2015 and 2016 respectively.

Simulation of Zero and Reduced Data

Once the data was prepared, by initially running the two stages of the validation and interpolation process, counting data was altered from the dataset to simulate anomalies in the bicycle counting data. For each bike season, 25 data anomalies were randomly created. For each anomaly, a counting site and date during the cycling season were randomly selected. Then, an anomaly duration was randomly chosen from the following possibilities: 6 hours (7:00 a.m. to 1:00 p.m.), 12 hours (7:00 a.m. to 7:00 p.m.), 1 day, 2 days, and 5 days. For the Arlington data, a total of 64 days of data anomalies were randomly selected for 2015, of which four were 6 hours' duration, five were 12 hours and 55 were full-day anomalies. In 2016, a total of 45 days of data anomalies were randomly selected, of which six were 6 hours, six were 12 hours and 33 were full days.

Three separate trials were performed. In the first trial, anomalies were created by replacing ground truth hourly counts with zero values for the duration of each anomaly. This situation represents either a complete blockage of the bicycle facility, or an automatic counter going offline. In the second and third trials, the hourly counts were reduced to 25% and 40% of the ground truth values, respectively, for the same randomly chosen counters, dates, and durations. The second and third trials represent situations that diminish but do not reduce bicycle counts to zero, such as construction projects near bicycle facilities, or partial blockages of a bicycle facility at the counting site location.

Results and Discussion

The data validation process is highly effective at filtering out days with significantly reduced bicycle count anomalies. Table 2 shows a summary, by counting site and by year, of all the days reduced to zero and 25% of the measured bicycle counts, and the proportion of those days that were subsequently flagged and interpolated. When combining the 2 years of analysis, 113 counter-days were simulated to have anomalous data in Arlington, of which, 98 days were flagged and interpolated in both the first (zero data) and second (25% of the measured bicycle counts) trials. None of the 10 days with only 6 hours of anomalous data were flagged, 10 out of 11 days with 12 hours of anomalous data were flagged and interpolated, and 88 out of 92 days with the entire day of anomalous data were flagged and interpolated. The days with 6 hours of anomalous data were not flagged because the daily factor was not decreased sufficiently to be triggered in Step 5 of the validation procedure. Had a lower *e* value been used, more of these days would have been flagged; however, it would have come at the cost of false positives due to inherent variation in daily factors between sites. Most (91%) of the days with 12 hours of anomalous data were flagged and interpolated because they had a more substantial decrease in their daily factor. Almost all (96%) of the full days of anomalous data were flagged and interpolated. There were four full days of anomalous data that were not flagged (at the Roosevelt Bridge counting site in 2015) because one of the two reference sites had anomalous data over the same period. This result exposes one of the vulnerabilities of the process in detecting zero or reduced data; if multiple counting sites have periods of zero or reduced data simultaneously, the ability of

Table 2. Results from Automated Validation of Data Anomaly Simulations

	2015				2016			
		Anomalies/flagged				Anomalies/flagged		
Arlington counting site	AADB	6 h	12 h	Full day	AADB	6 h	12 h	Full day
Trials 1 and 2: reduced to 0% and 25% of ground truth								
14th Street Bridge	1739			10/10	1896			5/5
Arlington Mill Drive Trail	350		NA		354			7/7
Ballston Connector	288			14/14	291			5/5
Bikeometer	1246			5/5	1302			3/3
Bluemont Connector	199			7/7	208		NA	
CC Connector	689		1/1		722	0 /1		
Custis Bon Air	1009				1026		1/1	4/4
Custis Rosslyn	1321		1/1		1434	0/1		
Key Bridge	1955				1943	0/1		
MVT Airport South	2155				2125		2/2	
Roosevelt Bridge	398			5/9	445		1/1	2/2
TR Island	1332	0/1			1308	0/1		
WOD Bon Air East	1014	0/1	0/1	3/3	968	0/1	1/1	1/1
WOD Bon Air West	1504	0/1	1/1	5/5	1484			
WOD Columbia Pike	846	0/1	1/1	6/6	805	0/1	1/1	6/6
WOD East Falls Church	750		NA		1116			
Total		0/4	4/5	55/59		0/6	6/6	33/33
Trial 3: reduced to 40% of ground truth								
14th Street Bridge	1739			10/10	1896			5/5
Arlington Mill Drive Trail	350		NA		354			7/7
Ballston Connector	288			14/14	291			4/5
Bikeometer	1246			5/5	1302			1/3
Bluemont Connector	199			7/7	208		NA	
CC Connector	689		1/1		722	0/1		
Custis Bon Air	1009				1026		0/1	4/4
Custis Rosslyn	1321		0/1		1434	0/1		
Key Bridge	1955				1943	0/1		
MVT Airport South	2155				2125		0/2	
Roosevelt Bridge	398			5/9	445		1/1	2/2
TR Island	1332	0/1			1308	0/1		
WOD Bon Air East	1014	0/1	0/1	3/3	968	0/1	1/1	1/1
WOD Bon Air West	1504	0/1	0/1	5/5	1484			
WOD Columbia Pike	846	0/1	1/1	5/6	805	0/1	1/1	6/6
WOD East Falls Church	750		NA		1116			
Total		0/4	2/5	54/59		0/6	3/6	30/33

Note: NA = not available.

the process to identify those anomalies is diminished. However, a well-maintained set of long-term counters is not likely to have multiple counting sites offline over the same period. The trials in this study introduced considerably more anomalies to the set of counting sites than what was typically observed in the bicycle counting data.

The results from the third trial, in which bicycle counts were reduced to 40% of the measured bicycle counts for the duration of the simulated anomaly, are also presented in Table 2. The results indicate that an anomaly causing bicycle counts to reduce to 40% of ground truth counts is likely to be detected by the validation program only if the anomaly persists for most of the day. A total of 89 days of anomalous data were flagged, compared with 98 days in the first and second trials. None of the 10 days with only 6 hours of anomalous data were flagged, which was unchanged from the first and second trials. The process did not perform as well in cases with 12 hours of anomalous data in the third trial compared with the first two; only 5 out of 11 days were flagged and interpolated. Although most (84 out of 92 days) anomalies spanning an entire day were flagged and interpolated, the results were inferior when compared with the first two trials. Four full days of anomalous data were not flagged by the process that had been flagged in the first two trials. These results are in concert with the theoretical limits of the process. If the parameter e is set at 2, as it was in all three trials, then the filtering process is not likely to detect anomalies that reduce bicycle counts by less than half of the daily total.

In cases where the anomaly was flagged, a daily factor was then interpolated by averaging the daily factor of the two most similar reference counting sites. Table 3 presents the anomalous daily factors for all three trials, the interpolated daily factors in the first (zero-data) trial and the ground truth factors for each of the 98 days that were flagged in the first and second trials. On average, the interpolated daily factor in the first and second trial had an ARE of 9.5% when compared with the ground truth; the third trial had a ARE of 10.2%. Note that 9 days have missing anomalous daily factors in the third trial because those days were not flagged in that trial. The distribution of ARE values is shown in Figure 2, illustrating that most of the 98 interpolated daily factors had an ARE <10%. For illustrative purposes, the daily factors for the counting site known as Ballston Connector, for 2015 are shown in Figure 3. The green trendline shows the ground truth daily factors. The red triangles represent the 14 days of anomalies simulated in the first trial, blue squares in the second trial, and orange circles in the third trial. The black diamonds represent the interpolated daily factors for the same 14 days in the first trial. The interpolated daily factors follow the ground truth factors closely. The ARE of the 14 days ranged from 1% to 16%. Note that the interpolated daily factors varied slightly when comparing the first two trials with the third trial. In several cases, a counting site with multiple days of anomalies would be assigned a different reference in the third trial when compared with the first two trials, which could result in slightly differing interpolated daily factors. To conserve space, the interpolated daily factors in figures 2 and 3 and Table 3 are only included for the first trial.

Conclusion

The accuracy of AADB estimation from short-term counts is highly sensitive to error in the reference data, particularly when the reference experiences significant undercounting during the short-term count period. An automated validation and interpolation process for reference bicycle count data would improve the quality of the reference data and, in the process, make AADB estimation from short-term counts more feasible in practice. This study proposed a methodology for grouping similar counting sites, validating the data by comparing daily factors of similar sites and interpolating any periods of anomalous data using the daily factors of the similar counting sites. The methodology was then tested on long-term bicycle count data from Arlington, with simulated anomalies. In three trials, data anomalies were artificially created that reduced counts to 0%, 25%, and 40%, respectively, of the measured bicycle counts for 6 consecutive hours, 12 consecutive hours, a full day, two consecutive full days, and five consecutive full days.

The days with 6 hours of anomalous data were not flagged in any of the three trials because the daily factors were not decreased sufficiently to be identified in the validation procedure. In the first and second trials, 91% of the days with 12 hours of anomalous data were flagged compared with 45% in the third trial. Similarly, in the first and second trials, 96% of full-day anomalies were flagged compared with 91% in the third trial. The third trial, in which hourly bicycle counts were reduced to 40% of measured values, approached the validation program limit. Given the parameters used in the study, if an anomaly does not reduce counts to below 50% of the ground truth daily total, then the validation program is not likely to flag the anomaly. When anomalies are flagged, the interpolated daily factors are close to the ground truth. Of the 98 anomalies flagged in the first and second trials, the AARE of the interpolated daily factor was 9.5%. Of the 89 anomalies flagged in the third trial, the AARE was 10.2%.

The proposed methodology, allowing for automatic validation and interpolation of reference bicycle count data, is part of a research project conducted in partnership with Eco-Counter. The methodology has been incorporated into software that will be accessible to jurisdictions with Eco-Counter equipment. However, the proposed methodology could be applied to continuous daily bicycle count data collected from any bicycle counting system and could be incorporated into other bicycle data archive and visualization tools. By applying this methodology, a transportation agency with multiple proximate reference counters could identify missing and erroneous data and subsequently interpolate to correct it. Two limitations of the proposed methodology must be highlighted. Firstly, an absolute minimum of three proximate reference counters are needed, thus this methodology is most applicable to bicycle

Table 3. Reduced, Interpolated and Ground Truth Daily Factors for Three Data Anomaly Trials

Date	Counting site	Daily factor for zero anomaly Trial 1	Daily factor for 25% anomaly Trial 2	Daily factor for 40% anomaly Trial 3	Interpolated factor for zero anomaly Trial 1	Factor (ground truth)	ARE (%) for zero anomaly Trial 1
15-07-16	14th Street Bridge	0.000	0.410	0.652	1.467	1.588	8%
15-07-17	14th Street Bridge	0.000	0.310	0.489	1.169	1.194	2%
15-07-18	14th Street Bridge	0.000	0.180	0.282	0.832	0.691	21%
15-07-19	14th Street Bridge	0.000	0.200	0.315	0.835	0.768	9%
15-07-20	14th Street Bridge	0.000	0.296	0.468	1.045	1.141	8%
15-09-10	14th Street Bridge	0.000	0.201	0.320	0.755	0.777	3%
15-09-11	14th Street Bridge	0.000	0.341	0.540	1.350	1.315	3%
15-09-12	14th Street Bridge	0.000	0.067	0.106	0.563	0.259	117%
15-09-13	14th Street Bridge	0.000	0.248	0.391	1.205	0.956	26%
15-09-14	14th Street Bridge	0.000	0.345	0.544	1.345	1.331	1%
15-04-02	Ballston Connector	0.000	0.314	0.471	1.044	1.144	9%
15-06-21	Ballston Connector	0.000	0.245	0.370	1.000	0.894	12%
15-08-01	Ballston Connector	0.000	0.300	0.464	1.133	1.106	2%
15-08-02	Ballston Connector	0.000	0.260	0.406	1.134	0.988	15%
15-08-03	Ballston Connector	0.000	0.311	0.482	1.160	1.158	0%
15-08-04	Ballston Connector	0.000	0.340	0.529	1.221	1.270	4%
15-08-05	Ballston Connector	0.000	0.322	0.496	1.247	1.207	3%
15-09-06	Ballston Connector	0.000	0.274	0.420	1.198	1.005	19%
15-09-07	Ballston Connector	0.000	0.333	0.514	1.151	1.249	8%
15-09-08	Ballston Connector	0.000	0.318	0.493	1.198	1.190	1%
15-09-09	Ballston Connector	0.000	0.311	0.486	1.273	1.151	11%
15-09-10	Ballston Connector	0.000	0.234	0.359	0.843	0.859	2%
15-11-23	Ballston Connector	0.000	0.179	0.279	0.691	0.685	1%
15-11-24	Ballston Connector	0.000	0.190	0.290	0.712	0.692	3%
15-04-17	Bluemont Connect.	0.000	0.247	0.385	1.013	0.919	10%
15-04-18	Bluemont Connect.	0.000	0.361	0.544	1.341	1.331	1%
15-07-12	Bluemont Connect.	0.000	0.340	0.523	1.205	1.306	8%
15-07-13	Bluemont Connect.	0.000	0.247	0.359	0.844	0.879	4%
15-07-14	Bluemont Connect.	0.000	0.309	0.477	1.129	1.180	4%
15-07-15	Bluemont Connect.	0.000	0.304	0.472	1.158	1.145	1%
15-07-16	Bluemont Connect.	0.000	0.381	0.605	1.428	1.512	6%
15-09-30	CC Connector	0.128	0.285	0.372	0.837	0.742	13%
15-07-10	Custis Rosslyn	0.190	0.461		1.269	1.266	0%
15-04-21	Roosevelt Bridge	0.000	0.356	0.562	1.258	1.368	8%
15-04-22	Roosevelt Bridge	0.000	0.256	0.390	0.863	0.969	11%
15-05-11	Roosevelt Bridge	0.000	0.418	0.655	1.390	1.611	14%
15-05-12	Roosevelt Bridge	0.000	0.385	0.606	1.414	1.481	4%
15-11-17	Roosevelt Bridge	0.000	0.279	0.431	0.988	1.064	7%
15-04-04	WOD Bon Air East	0.000	0.148	0.233	0.540	0.577	6%
15-04-05	WOD Bon Air East	0.000	0.339	0.536	1.150	1.333	14%
15-08-21	WOD Bon Air East	0.000	0.302	0.481	1.220	1.194	2%
15-04-07	WOD Bon Air West	0.000	0.170	0.267	0.645	0.663	3%
15-04-08	WOD Bon Air West	0.000	0.109	0.171	0.428	0.423	1%
15-04-09	WOD Bon Air West	0.000	0.096	0.150	0.440	0.373	18%
15-04-10	WOD Bon Air West	0.000	0.102	0.159	0.433	0.396	9%
15-04-11	WOD Bon Air West	0.000	0.348	0.555	1.358	1.373	1%
15-11-17	WOD Bon Air West	0.096	0.246		0.778	0.681	14%
15-06-30	WOD Columbia Pike	0.000	0.260	0.410	1.113	1.009	10%
15-07-01	WOD Columbia Pike	0.000	0.236	0.371	0.993	0.912	9%
15-07-02	WOD Columbia Pike	0.000	0.173	0.272	0.694	0.672	3%

(continued)

Table 3. (continued)

Date	Counting site	Daily factor for zero anomaly Trial 1	Daily factor for 25% anomaly Trial 2	Daily factor for 40% anomaly Trial 3	Interpolated factor for zero anomaly Trial 1	Factor (ground truth)	ARE (%) for zero anomaly Trial 1
15-07-03	WOD Columbia Pike	0.000	0.470		1.800	1.823	1%
15-07-04	WOD Columbia Pike	0.000	0.167	0.259	0.748	0.639	17%
15-07-11	WOD Columbia Pike	0.000	0.375	0.593	1.373	1.458	6%
15-09-22	WOD Columbia Pike	0.113	0.315	0.434	0.990	0.902	10%
15-11-18	Bikeometer	0.000	0.209	0.328	0.934	0.815	15%
15-11-19	Bikeometer	0.000	0.089	0.135	0.436	0.339	29%
15-11-20	Bikeometer	0.000	0.204	0.322	0.827	0.797	4%
15-11-21	Bikeometer	0.000	0.133	0.206	0.525	0.513	2%
15-11-22	Bikeometer	0.000	0.065	0.103	0.332	0.258	29%
16-05-03	14th Street Bridge	0.000	0.153	0.243	0.645	0.604	7%
16-05-04	14th Street Bridge	0.000	0.201	0.316	0.774	0.786	2%
16-05-05	14th Street Bridge	0.000	0.184	0.289	0.708	0.717	1%
16-05-06	14th Street Bridge	0.000	0.055	0.087	0.262	0.216	21%
16-05-07	14th Street Bridge	0.000	0.152	0.241	0.663	0.600	11%
16-09-21	Arlington Mill Drive	0.000	0.304	0.481	1.207	1.173	3%
16-09-22	Arlington Mill Drive	0.000	0.273	0.421	1.279	1.049	22%
16-10-08	Arlington Mill Drive	0.000	0.045	0.063	0.15	0.180	17%
16-10-09	Arlington Mill Drive	0.000	0.149	0.237	0.684	0.581	18%
16-10-10	Arlington Mill Drive	0.000	0.264	0.409	1.034	1.029	0%
16-10-11	Arlington Mill Drive	0.000	0.215	0.335	0.897	0.840	7%
16-10-12	Arlington Mill Drive	0.000	0.270	0.418	1.059	1.015	4%
16-09-02	Ballston Connector	0.000	0.319	0.500	1.247	1.239	1%
16-09-03	Ballston Connector	0.000	0.249		0.874	0.933	6%
16-09-04	Ballston Connector	0.000	0.302	0.479	1	1.191	16%
16-09-05	Ballston Connector	0.000	0.302	0.465	1.157	1.160	0%
16-09-06	Ballston Connector	0.000	0.344	0.535	1.432	1.321	8%
16-05-13	Custis Bon Air	0.000	0.165	0.262	0.651	0.649	0%
16-05-14	Custis Bon Air	0.000	0.243	0.385	1.022	0.952	7%
16-09-20	Custis Bon Air	0.183	0.468		1.198	1.289	7%
16-11-02	Custis Bon Air	0.000	0.323	0.510	1.154	1.262	9%
16-11-03	Custis Bon Air	0.000	0.218	0.343	0.797	0.851	6%
16-07-09	MVT Airport South	0.146	0.452		1.285	1.359	5%
16-08-05	MVT Airport South	0.198	0.418		1.095	1.068	3%
16-10-27	Roosevelt Bridge	0.000	0.213	0.328	0.737	0.820	10%
16-10-28	Roosevelt Bridge	0.000	0.251	0.384	0.887	0.964	8%
16-11-26	Roosevelt Bridge	0.013	0.080	0.117	0.267	0.272	2%
16-05-17	WOD Bon Air East	0.056	0.087	0.099	0.227	0.165	37%
16-09-02	WOD Bon Air East	0.000	0.378	0.601	1.448	1.495	3%
16-04-03	WOD Columbia Pike	0.000	0.100	0.160	0.363	0.393	8%
16-04-04	WOD Columbia Pike	0.000	0.103	0.158	0.437	0.383	14%
16-04-05	WOD Columbia Pike	0.000	0.069	0.104	0.307	0.264	16%
16-04-06	WOD Columbia Pike	0.000	0.129	0.198	0.497	0.494	1%
16-04-07	WOD Columbia Pike	0.000	0.054	0.078	0.169	0.198	15%
16-09-30	WOD Columbia Pike	0.000	0.100	0.158	0.429	0.392	9%
16-11-18	WOD Columbia Pike	0.068	0.249	0.354	0.836	0.775	8%
16-04-17	Bikeometer	0.000	0.296		1.059	1.169	9%
16-04-18	Bikeometer	0.000	0.324		1.031	1.280	19%
16-09-28	Bikeometer	0.000	0.144	0.228	0.609	0.567	7%

Note: ARE = absolute relative error.

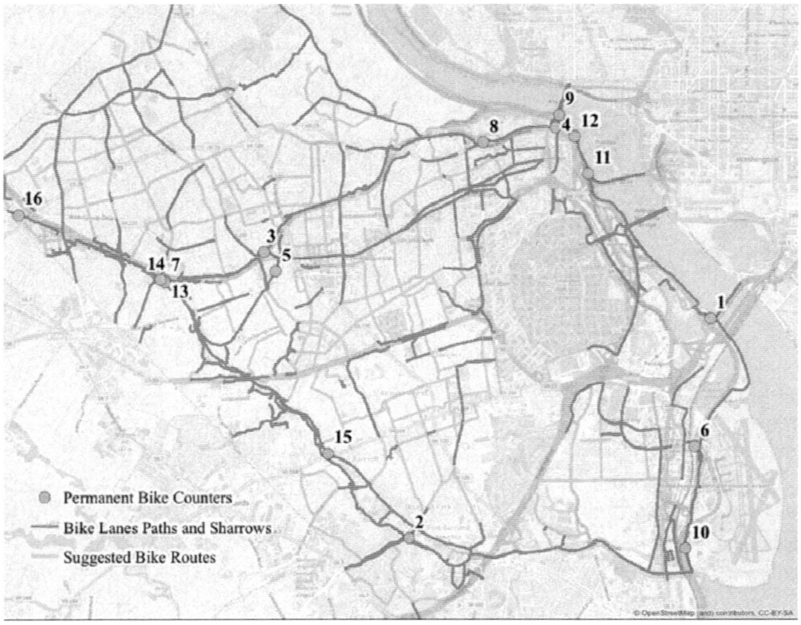

Figure 1. Long-term bicycle counters in Arlington, VA: 14th Street Bridge (1), Arlington Mill Drive Trail (2), Ballston Connector (3), Bikeometer (4), Bluemont Connector (5), CC Connector (6), Custis Bon Air (7), Custis Rosslyn (8), Key Bridge (9), MVT Airport South (10), Roosevelt Bridge (11), TR Island (12), WOD Bon Air East (13), WOD Bon Air West (14), WOD Columbia Pike (15), and WOD East Falls Church (16).

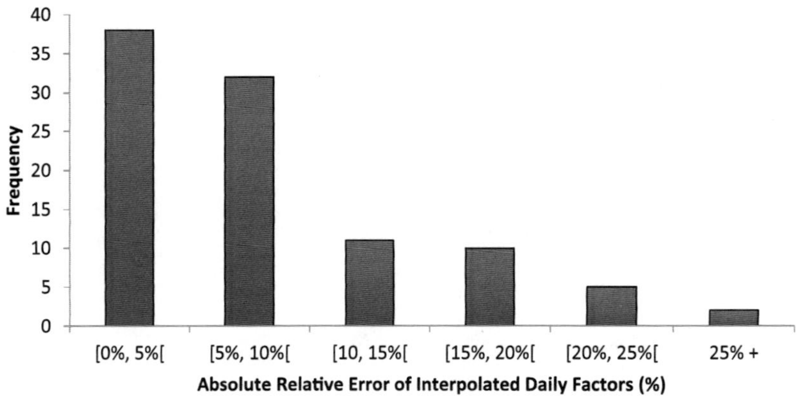

Figure 2. Frequency of ARE of interpolated daily factors for first trial, zero-anomaly simulations.

counting sites across an urban center or region. Ideally, a transportation agency might have 10 or more reference counters across their jurisdiction to collect bicycle count data at several similar locations for each type of cycling facility present across the bicycle network. Secondly, in colder climates, this analysis is only appropriate during the cycling season. In this study, the cycling season was defined as all months with at least 50% of the cycling demand compared with the peak month. During the winter season, in many jurisdictions located in cold climatic zones, a combination of low cycling demand, high weather variability, and great variability in surface conditions across cycling facilities complicate the validation and interpolation process making an automated tool prone to error.

Future work will include testing the validation and interpolation tool developed in this paper with larger datasets and testing anomalies that result in increased bicycle counts. The logic behind the validation process could also be expanded to allow for more than two references to be used. The tool is intended to be the first stage of a larger program that estimates AADB from short-term counts. The program would receive both short-term

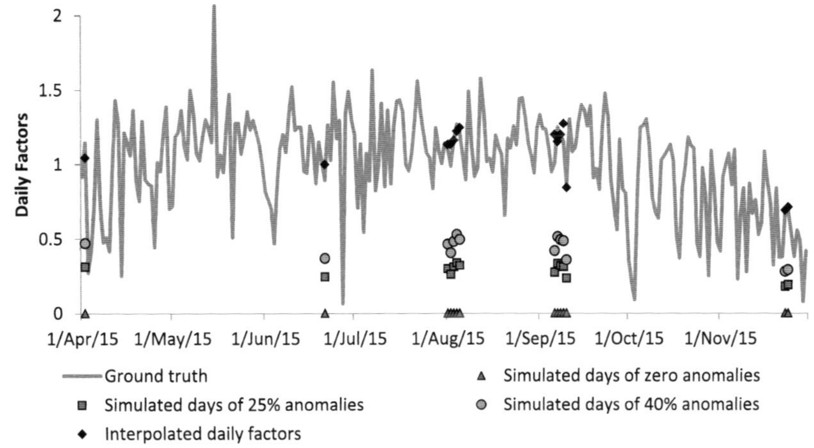

Figure 3. Example of interpolated daily factors after the automated data validation and interpolation process.

counts and long-term reference counts as inputs. The program would then validate and interpolate reference data, cluster validated reference sites into similar groups, match short-term counts to reference clusters, estimate the quality of the short-term count, and finally estimate the AADB of the short-term counting site. By completely automating the process of data validation and AADB extrapolation of short-term counts, transportation agencies would find it more feasible to conduct comprehensive bicycle count programs that extend to hundreds of locations through a city or regional bike network. The outcome would help practitioners in bicycle infrastructure planning, and in evaluating injury risk across the network.

Acknowledgments

Funding for this project was provided in part by the Natural Sciences and Engineering Research Council and Eco-Counter. We thank Arlington County for access to their bicycle counting data.

Author Contributions

The authors confirm contribution to the paper as follows: study conception and design: DB; data collection: Eco-Counter, Arlington County; analysis and interpretation of results: DB, SM, FM, LM-M; draft manuscript preparation: DB. All authors reviewed the results and approved the final version of the manuscript.

References

1. Fournier, N., E. Christofa, and M. A. Knodler. A Mixed Methods Investigation of Bicycle Exposure in Crash Rates. *Accident Analysis & Prevention*, March 1, 2017. Advance online publication.
2. Nosal, T., and L. F. Miranda-Moreno. Cycling and Weather: A Multi-City and Multi-Facility Study in North America. Presented at 91st Annual Meeting of theTransportation Research Board, Washington, D.C., 2012.
3. Nordback, K., W. E. Marshall, B. N. Janson, and E. Stolz. Estimating Annual Average Daily Bicyclists: Error and Accuracy. *Transportation Research Record: Journal of the Transportation Research Board*, 2013. 2339: 90–97.
4. Nosal, T., and L. F. Miranda-Moreno. The Effect of Weather on the Use of North American Bicycle Facilities: A Multi-City Analysis Using Automatic Counts. *Transportation Research Part A: Policy and Practice*, Vol. 66, No. 1, 2014. 213–225.
5. Beitel, D., S. McNee, and L. F. Miranda-Moreno. Quality Measure of Short-Duration Bicycle Counts. *Transportation Research Record: Journal of the Transportation Research Board*, 2017. 2644: 64–71.
6. Beitel, D., and L. Miranda-Moreno. Methods for Improving and Automating the Estimation of Average Annual Daily Bicyclists. Presented at 95th Annual Meeting of the Transportation Research Board, Washington, D.C., 2016.
7. Miranda-Moreno, L. F., T. Nosal, R. J. Schneider, and F. Proulx. Classification of Bicycle Traffic Patterns in Five North American Cities. *Transportation Research Record: Journal of the Transportation Research Board*, 2013. 2339: 68–79.
8. Nosal, T., L. F. Miranda-Moreno, and Z. Krstulic. Incorporating Weather: A Comparative Analysis of Average Annual Daily Bicyclist Estimation Methods. Presented at 93rd Annual Meeting of the Transportation Research Board, Washington, D.C., 2014.
9. Hankey, S., G. Lindsey, and J. Marshall. Day-of-Year Scaling Factors and Design Considerations for Nonmotorized Traffic Monitoring Programs. *Transportation Research Record: Journal of the Transportation Research Board*, 2014. 2468: 64–73.
10. Figliozzi, M., P. Johnson, C. Monsere, and K. Nordback. Methodology to Characterize Ideal Short-Term Counting Conditions and Improve AADT Estimation Accuracy Using a Regression-Based Correcting Function. *Journal of Transportation Engineering*, Vol. 140, No. 5, 2014, p. 04014014.
11. El Esawey, M., and A. I. Mosa. Determination and Application of Standard *K* Factors for Bicycle Traffic. *Transportation Research Record: Journal of the Transportation Research Board*, 2015. 2527: 58–68.
12. El Esawey, M., C. Lim, T. Sayed, and A. I. Mosa. Development of Daily Adjustment Factors for Bicycle Traffic. *Journal of Transportation Engineering*, Vol. 139, No. 8, 2013. 859–871.

13. El Esawey, M. Estimation of Annual Average Daily Bicycle Traffic with Adjustment Factors. *Transportation Research Record: Journal of the Transportation Research Board*, 2014. 2443: 106–114.
14. El Esawey. Toward a Better Estimation of Annual Average Daily Bicycle Traffic: Comparison of Methods for Calculating Daily Adjustment Factors. *Transportation Research Record: Journal of the Transportation Research Board*, 2016. 2593: 28–36.
15. Nosal, T. *Improving the Accuracy of Bicycle AADT Estimation: Temporal Patterns, Weather and Bicycle AADT Estimation Methods*. Masters degree thesis. Civil Engineering and Applied Mechanics, McGill University, Montreal, 2014.
16. Fournier, N., E. Christofa, and M. A. Knodler. A Sinusoidal Model for Seasonal Bicycle Demand Estimation. *Transportation Research Part D: Transport and Environment*, Vol. 50, 2017. 154–169.
17. Lindsey, G. *The Minnesota Bicycle and Pedestrian Counting Initiative: Implementation Study*. Minnestoa Department of Transportation, St. Paul, MN, 2015.
18. Jackson, K. N., E. Stolz, and C. Cunningham. Nonmotorized Site Selection Methods for Continuous and Short-Duration Volume Counting. *Transportation Research Record: Journal of the Transportation Research Board*, 2015. 2527: 49–57.
19. FHWA. *Traffic Monitoring Guide*. U.S. Department of Transportation, Federal Highway Administration, Washington, D.C., 2013.
20. Jackson, K. N., S. W. O'Brien, S. E. Searcy, and S. E. Warchol. Quality Assurance and Quality Control Processes for a Large-Scale Bicycle and Pedestrian Volume Data Program. *Transportation Research Record: Journal of the Transportation Research Board*, 2017. 2644: 19–29.
21. TRB. *Guidebook on Pedestrian and Bicycle Volume Data Collection*. The National Academies Press, Washington, D.C., 2014.
22. Nordback, K., K. Tufte, N. McNeil, M. Harvey, and M. Watkins. *Bike-Ped Portal: Development of an Online Nonmotorized Traffic Count Archive*. National Institute for Transportation and Communities (NITC), Portland, OR, 2017.
23. Miranda-Moreno, L., and T. Nosal. Weather or Not to Cycle. *Transportation Research Record: Journal of the Transportation Research Board*, 2011. 2247: 42–52.
24. Nordback, K., D. Piatkowski, B. N. Janson, W. E. Marshall, K. J. Krizek, and D. S. Main. Using Inductive Loops to Count Bicycles in Mixed Traffic. *Journal of Transportation of the Institute of Transportation Engineers*, Vol. 2, No. 1. 39–56, 2011.
25. Nordback, K., and B. Janson. Automated Bicycle Counts: Lessons from Boulder, Colorado. *Transportation Research Record: Journal of the Transportation Research Board*, 2010. 2190: 11–18.

The Standing Committee on Highway Traffic Monitoring (ABJ35) peer-reviewed this paper (18-04717).

TRR
JOURNAL OF THE TRANSPORTATION RESEARCH BOARD

Article

Transportation Research Record
2018, Vol. 2672(43) 87–97

DOI: 10.1177/0361198118793006
journals.sagepub.com/home/trr
SAGE

Comparison of Pedestrian Count Expansion Methods: Land Use Groups versus Empirical Clusters

Julia B. Griswold[1], Aditya Medury[1], Robert J. Schneider[2], and Offer Grembek[1]

Abstract

Expansion factors based on the trends in long-term count data are useful tools for estimating daily, weekly, or annual volumes from short-term counts, but it is unclear how to differentiate locations by activity pattern. This paper compares two approaches to developing factor groups for hour-to-week pedestrian count expansion factors. The land use (LU) classification approach assumes that surrounding LUs affect the pedestrian activity at a location, and it is easy to apply to short-term count locations based on identifiable attributes of the site. The empirical clustering (EC) approach uses statistical methods to match locations based on the actual counts, which may produce more accurate volume estimates, but presents a challenge for determining which factor group to apply to a location. We found that both the LU and EC approaches provided better weekly pedestrian volume estimates than the single factor approach of taking the average of all locations. Further, the differences between LU and EC estimation errors were modest, so it may be beneficial to use the intuitive and practical LU approach. LU groupings can also be modified with insights from the EC results, thus improving estimates while maintaining the ease of application. Ideal times for short-term counts are during peak activity periods, as they generally produce estimates with fewer errors than off-peak periods. Weekly volume estimated from longer-duration counts (e.g., 12 h) is generally more accurate than estimates from shorter-duration counts (e.g., 2 h). Practitioners can follow this guidance to improve the quality of weekly pedestrian volume estimates.

Annual pedestrian volumes can be a valuable exposure measure for safety and risk analyses. However, government agencies do not have the resources to collect long-term pedestrian counts at all relevant locations; many jurisdictions are only able to collect short-term peak-period counts at high-priority locations. In addition, counts collected during different time periods are not comparable and are only relevant for analysis of crashes occurring during that time. A number of pedestrian and bicycle studies have used automated count data from a limited number of locations to develop expansion factors that can be used to expand short-term counts to long-term volume estimates (*1–8*). This process is based on the assumption that similar sites will have similar patterns in hourly, weekly, and monthly volume trends. Even if the magnitude of volumes is different, the proportional trends can be similar.

In this paper, we explore a practical solution to developing hour-to-week pedestrian count expansion factors. We base the factors on a dataset of 103 unique locations, each with at least one week of automated counts. An a priori assumption of this study is that weekly pedestrian activity patterns vary significantly according to land use (LU) and that locations with similar patterns should be grouped together for more accurate expansion of short-term counts (these similar patterns are often called "factor groups" in the traffic monitoring field) (*7*, *9*). For example, compare the average activity patterns of two locations with more than 1 year of data shown in Figure 1. The pattern at the central business district (CBD) location (Figure 1*a*) has unimodal peaks around lunch time and the weekend activity is much lower than weekdays. The pattern near the elementary school (Figure 1*b*) has three daily peaks on weekdays, roughly corresponding with drop-off and pick-up times for students.

[1]Safe Transportation Research and Education Center, University of California, Berkeley, CA
[2]Department of Urban Planning, University of Wisconsin-Milwaukee, Milwaukee, WI

Corresponding Author:
Address correspondence to Julia B. Griswold: juliagris@berkeley.edu

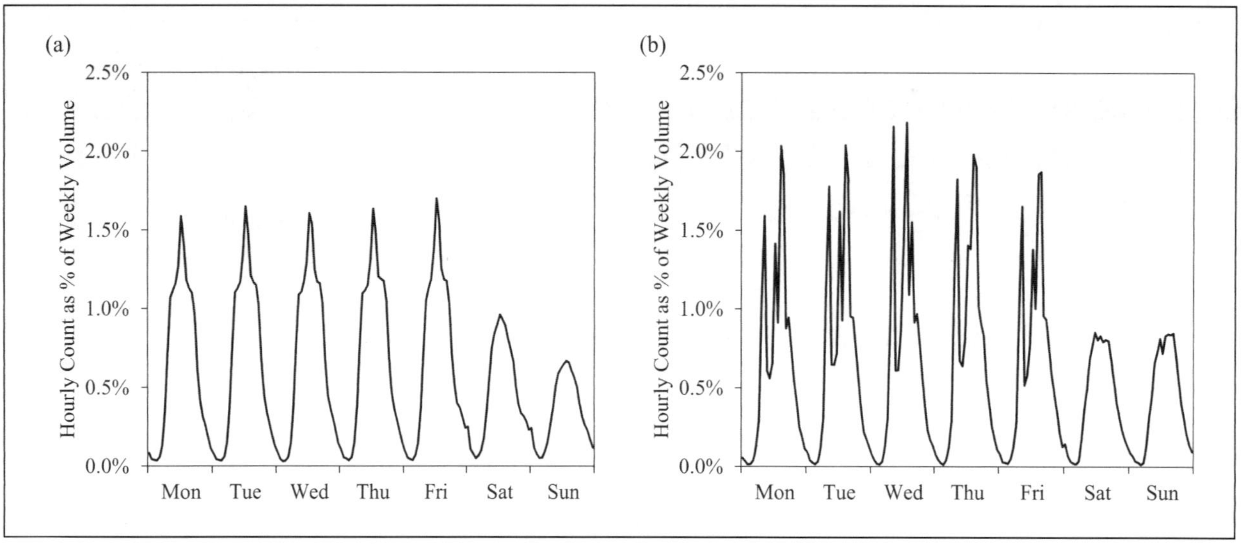

Figure 1. Weekly pedestrian activity patterns: (*a*) CBD; (*b*) elementary school.

The difference between these two patterns is important in practice. Imagine that an agency would like to estimate a weekly pedestrian volume at a downtown site based on a short-term count of 100 pedestrians between 8 and 9 a.m. on a Wednesday. Based on the CBD pattern, this time period represents approximately 1.1% of the weekly volume, expanding to a weekly estimate of 100/0.011 = 9,091. But if the agency incorrectly applies the elementary school pattern instead (this time period represents 2.2% of the weekly volume), its weekly estimate would be 100/0.022 = 4,545. The school expansion factor would underestimate the weekly volume by 50%.

Previous Approaches

Pedestrian and bicycle count expansion studies have generally used two main approaches to group long-term count patterns.

Land Use (LU) Classification Approach

Several researchers have grouped locations based upon the surrounding LU characteristics. This approach assumes that land uses are an important underlying cause of pedestrian activity, so locations near similar LUs will have similar pedestrian activity patterns throughout the day and week. Previous studies have categorized distinct pedestrian patterns for employment centers, residential areas, neighborhood commercial areas, and locations near multi-use trails (*1*, *10*) and categorized patterns based on employment density (*2*, *11*) as well as level of urbanism and climatic conditions (*12*). This approach is unlikely to produce pedestrian patterns that are as statistically distinct as the empirical method, but it can be more easily applied to short-term count locations based on identifiable attributes of the site.

Empirical Clustering (EC) Approach

A second approach is to identify similar types of activity patterns empirically, based on the count data itself. This method identifies clusters of similar patterns, seeking to minimize variation within each cluster and maximize variation between clusters. Colorado Department of Transportation used cluster analysis to identify three distinct day-to-week and three distinct month-to-year trail volume patterns (*4*). Miranda-Moreno et al. (*13*) also used a heuristic approach to manually classify bicycle activity patterns in five North American cities. They identified distinct patterns using two indices: 1) the ratio of weekend to weekday volume and 2) the ratio of morning (7 to 9 a.m.) to midday (11 a.m. to 1 p.m.) volume. This produced four distinct factor groups: utilitarian, mixed-utilitarian, mixed-recreational, and recreational. This approach has also been applied in other references (*5*, *7*), but not in a pedestrian-only setting. The empirical clustering (EC) approach is an effective way to match locations with similar patterns, but once those groups are created, it is unclear how to apply them to specific short-term count locations.

Despite research quantifying different types of pedestrian activity patterns, there is relatively little guidance about which approach to use (*7*, *9*). Further, existing guidance offers little help for determining which specific activity pattern should be applied to expand a count at a particular location. This paper compares the two approaches and uses the results to provide practical guidance for pedestrian count expansion.

Method

This section describes our long-term count data, defines hour-to-week expansion factors, and summarizes how we classified weekly pedestrian activity patterns into groups using both LU and EC approaches.

Long-Term Count Data

We compiled an initial dataset of 133 unique locations with count durations varying between 1 week and 5 years. Automated passive infrared counters from the same vendor were set up at sidewalk and trail locations in Alameda, Fresno, Los Angeles, San Diego, and San Francisco counties in California between 2008 and 2017 as part of unrelated research and planning programs. These counters collect screenline volumes and cannot capture intersection crossing volumes, but when set up on a sidewalk near an intersection, they likely capture similar activity patterns to the crosswalks. The counters aggregated counts to 15-min or 1-h intervals, and we analyzed all counts at 1-h intervals.

Although our study focuses on pedestrian count expansion factors, the passive infrared counters would have also counted bicyclists. We do not know the percentage of bicyclists counted at each site, but the majority of sites were sidewalk counts, and they (and the activity patterns they produced) were likely to be dominated by pedestrians. Bicyclist counts would only affect our results if there was a large proportion of bicyclists at a count site and the weekly activity pattern of these bicyclists was significantly different from the weekly pedestrian activity pattern.

For each location, we graphed the count values over time and flagged observations that were outliers. These outliers included intervals with spikes in counts that were more than double the values at corresponding times for surrounding weeks and extended periods of zero counts, both indicating counter errors or aberrations from normal activity (e.g., athletic event, parade, street festival). We also excluded national holidays. Following this cleanup, we dropped 30 of the locations that lacked at least one observation for each of the 168 h in a week. The remaining count locations are shown in Figure 2, and although not selected randomly, cover many different types of urban, suburban, and even some rural landscapes.

We focused our analysis on hour-to-week expansion factors. These are particularly important because they capture notable differences in activity due to nearby LUs. It is beyond our scope to analyze monthly or seasonal variation. However, in doing our analysis, we also provide a framework for developing expansion factors to account for seasonal variations and estimate annual volumes.

Hour-to-Week Expansion Factors

For each of the 103 (N) remaining sites, we calculated individual expansion factors by adapting the method described in Ryus et al. (*7*). Let $v_{i,t,d,w}$ correspond to the pedestrian count for a location i, hour of day t, day of week d, and week w. Given that the pedestrian counts for each location were collected for a varying number of weeks, an average weekly count is represented as follows:

$$\bar{v}_{i,t,d} = \frac{\sum_{w=1}^{W_{i,t,d}} v_{i,t,d,w}}{W_{i,t,d}}, \tag{1}$$

where $W_{i,t,d}$ represents the number of weeks for which count data is available for the given hour, day, and location combination. Once we estimate the average weekly pedestrian activity for a location, we can estimate a normalized count as follows:

$$\tilde{v}_{i,t,d} = \frac{\bar{v}_{i,t,d}}{\sum_{t=0}^{23}\sum_{d=0}^{6}\bar{v}_{i,t,d}} \tag{2}$$

where values associated with d range from 0 (Monday) to 6 (Sunday). Finally, we obtain a single expansion factor for a given time period t, and day d, $\gamma_{t,d}^{SF}$, as follows:

$$\gamma_{t,d}^{SF} = \frac{\sum_{i=1}^{N}\tilde{v}_{i,t,d}}{N}. \tag{3}$$

Figure 3 shows the mean, mean plus or minus the standard deviation, and the minimum and maximum hour-to-week expansion factors by time of week for all 103 count sites combined. The mean values represent a single set of hourly factors that would be used if we assumed that the activity patterns were the same across all sites and did not vary with LU. We call this the single factor (SF) activity pattern. Comparing to the trends in Figure 1, it is clear that the sites have different patterns. Note that the minimum (or maximum) curve does not represent a single site, but the set of lowest expansion factor values among all sites.

Land Use (LU) Classification Approach

We used Google Maps place data and satellite imagery to determine the surrounding LUs at each site. The LU categories and definitions are described in Table 1.

The development of the LU groupings was a two-stage process. First, we graphed the means of expansion

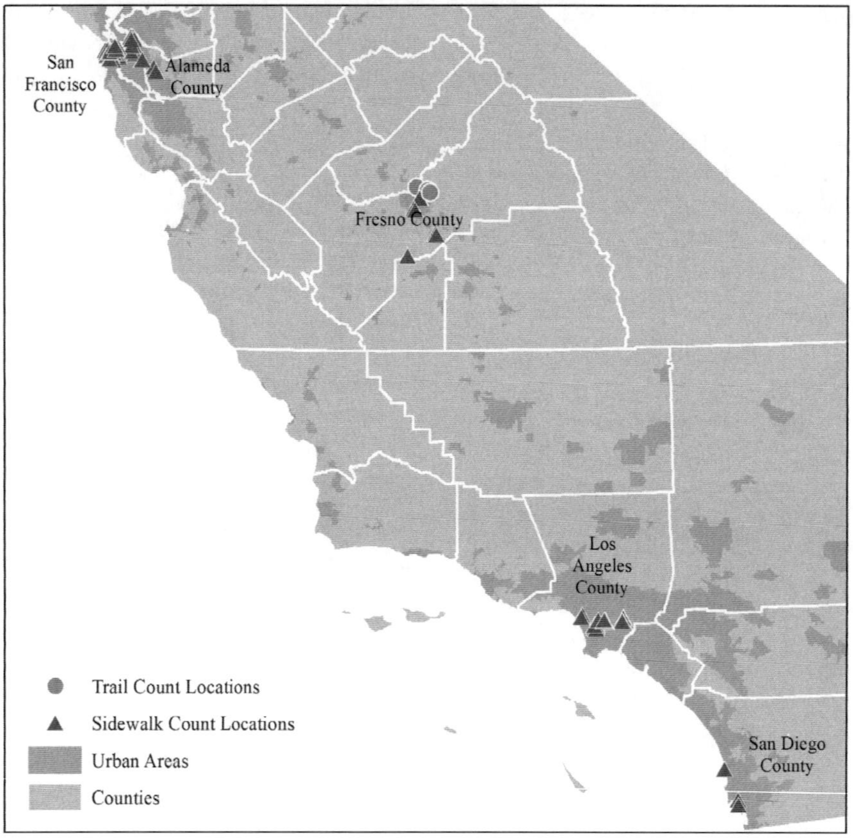

Figure 2. Count locations.

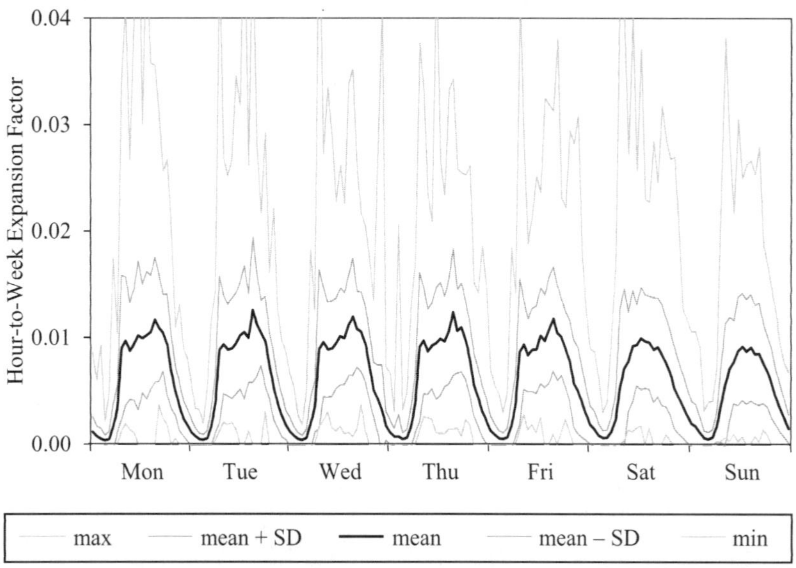

Figure 3. Single factor (SF) (103 sites) activity pattern: mean hourly values for all count sites.

Table 1. Land Use Definitions

Category	Definition
Central business district	In the downtown area as labeled on Google Maps or from expert knowledge of the area
School	School facility on adjacent block or yellow crosswalk present at intersection
Trail	Count location within a block of trail access point
Commercial	Two or more commercial (retail or office) businesses on adjacent blocks
Residential	Residential land use and no commercial on adjacent blocks
Park	Park on adjacent blocks
Rural	Agricultural fields on at least two adjacent blocks

factors for sites by LU category and judged which LUs had distinct patterns. We found that CBDs and schools had similar patterns to the corresponding graphs in Figure 1. Park, residential, and rural sites did not display meaningful trends, so these LU categories were grouped as an "Other" category. Subsequently, we created mutually exclusive groups, based on the following hierarchy:

1. CBD
2. School – excluding CBD locations
3. Trail – excluding CBD and school locations
4. Commercial – excluding CBD, school, and trail locations
5. Other – all remaining locations.

Empirical Clustering (EC) Approach

In parallel, we applied the EC approach, using the kml package in R to identify clusters with different trends. This package applies the k-means algorithm to longitudinal data with the objective of creating compact clusters that are well separated from each other (*14*). The EC approach requires setting a specific number of clusters prior to running the algorithm. We ran the analysis for one cluster (EC1, which is equivalent to SF) and two through six clusters (EC2 through EC6).

Two metrics were useful in evaluating the quality of the clustering output and determining the best number of clusters to choose:

- Sum of squared differences (SSD): This criterion sums the squared difference between a location's hour of week distribution and the mean distribution of the cluster to which the location corresponds.
- Calinski-Harabasz (CH) criterion (*15*): This criterion calculates the within-cluster and between-cluster variation to evaluate clustering quality. It can be interpreted as (between-cluster variation)/(within-cluster variation), thus a higher value implies more distinct clusters.

Although SSD was the primary criterion to identify the best cluster definition for each k, we used CH to capture how distinct the weekly patterns across clusters were.

The k-means approach follows a sequence of the following two steps until convergence is achieved:

- Maximization step: Using the weekly expansion factor that is associated with each cluster, each location is linked to a cluster number for which it has the smallest SSD value.
- Expectation step: Based on the locations attributed to each cluster, the weekly expansion factor for each cluster is re-estimated as the average of all locations that are associated with it.

Since the sequence of steps described above depends on the initial values of each cluster's expansion factors, we repeated the k-means approach for 800 iterations with randomized initial configuration, and stored the cluster definition with the lowest SSD for each k-value, or number of clusters.

Analysis

We classified all 103 count sites into mutually exclusive LU categories based on the hierarchy described above. Initially, 13 sites were classified as CBD, 23 as School, 6 as Trail, 38 as Commercial, and 23 as Other.

We also classified all of the sites into sets of clusters based on the EC approach. As the results of the clustering in Table 2 indicate, SSD decreased as the number of clusters increased (as expected). Moreover, the SSD corresponding to the LU-based classification is similar to the empirical clustering value corresponding to two clusters (EC2). However, the CH criterion is maximum for EC3. We selected EC4, the second best according to the CH criterion, because the activity patterns most closely matched the patterns that we expected from specific LU categories. The activity patterns and distributions for EC4 are shown in Figure 3.

Table 2. Quality Criteria for Empirical Clustering Results

Number of clusters	Sum of squared differences across all clusters	Calinski-Harabasz criterion
EC1	0.252	na
EC2	0.219	15.060
EC3	0.189	16.533
EC4	0.172	15.303
EC5	0.160	14.004
EC6	0.151	13.021
LU	0.199	6.462

The four clusters within EC4 match well with several LU activity patterns. Cluster 1 (Figure 4*a*) shows bimodal peaks in the morning and evening, suggesting commuting activity, along with greater activity levels on the weekend, suggesting recreational activity. These locations correspond exactly with the Trail sites (Figure 4*b*). Cluster 2 (Figure 4*c*), by far the largest cluster, with its slight peak in the afternoon, had a pattern resembling both the Commercial (Figure 4*d*) and Other (Figure 4*e*) LU groups. Cluster 3 (Figure 4*f*) resembled the School location in Figure 1*b* and checking the eight sites included in this cluster, all were within two blocks of a school. There were, however, 12 sites in the School LU category. Examining the sites near schools that ended up in Cluster 2 (commercial-like trend), we saw that most of them had retail or commercial uses on adjacent blocks. This result suggests that commercial activity may be a more dominating factor than school activity in creating the overall pedestrian activity pattern in an area. Cluster 4 (Figure 4*h*) was most similar to the CBD LU group with its midday peak, but the ratio between weekday and weekend volumes was different.

As a final step, based on the insights from examining the empirical clusters, we revised the hierarchy of LU groupings, so that sites with both school and commercial adjacent uses would be designated as Commercial LU. The updated LU classifications still did not produce as distinct groupings as the EC method according to our performance measures, but this was expected.

Comparison of Approaches

Next, we compared the LU and EC classification methods visually and quantitatively.

Visual Comparison

Visually, there is close correspondence between the LU and EC4 clusters (Figure 4). The Trail group is identical to Cluster 1. The School group has biomodal peaks and lower weekend activity like Cluster 3, although the peaks are not as high. The CBD group has similar weekday patterns to Cluster 4, but the weekend activity does not drop as significantly. The Commercial and Other groups both resemble Cluster 2. On average, sites near commercial areas tend to see increasing activity through weekday mornings as shops open and have the highest amounts of activity on weekday afternoons and early evenings. Weekend shopping area activity peaks during the middle of the day.

Cross-Tabulation of LU and EC Categories

We also compared the categories produced by the two classification methods. Our cross-tabulation in Table 3 shows that all six Trail sites were in empirical Cluster 1. Most of the Commercial and Other sites were in the expected Cluster 2, which captured 71% of the locations. Despite the visual similarities between their patterns, at least 50% of the CBD and School sites were in empirical clusters with different apparent trends. In fact, only 38% of the CBD sites were in Cluster 4, which seemed to have the most CBD-like pattern. The majority of CBD sites were in Cluster 2, which corresponded with Commercial and Other sites.

The results, especially for CBD sites, highlight the challenge of categorizing mixed-use locations. CBDs are generally characterized by a mix of LUs, but they can have very different characteristics between cities. Some CBDs are strictly government and office-type business centers, but others may also have entertainment and retail districts, which would also have activity on weekends and outside regular business hours. Given the differences between the cluster results, it is worthwhile examining the quality of the weekly estimates that would be produced from a single CBD expansion factor.

General Expansion Accuracy

For the comparison, we considered three cluster types:

- Single factor (SF), obtained by averaging the distributions of all locations
- Land use-based clusters (LU)
- Best fit empirical cluster (EC4)

For the purposes of comparing the expansion accuracies of different clustering approaches, the in-sample estimation error is defined as the absolute percentage error of expanded weekly counts relative to the average observed weekly counts. This metric is calculated as follows:

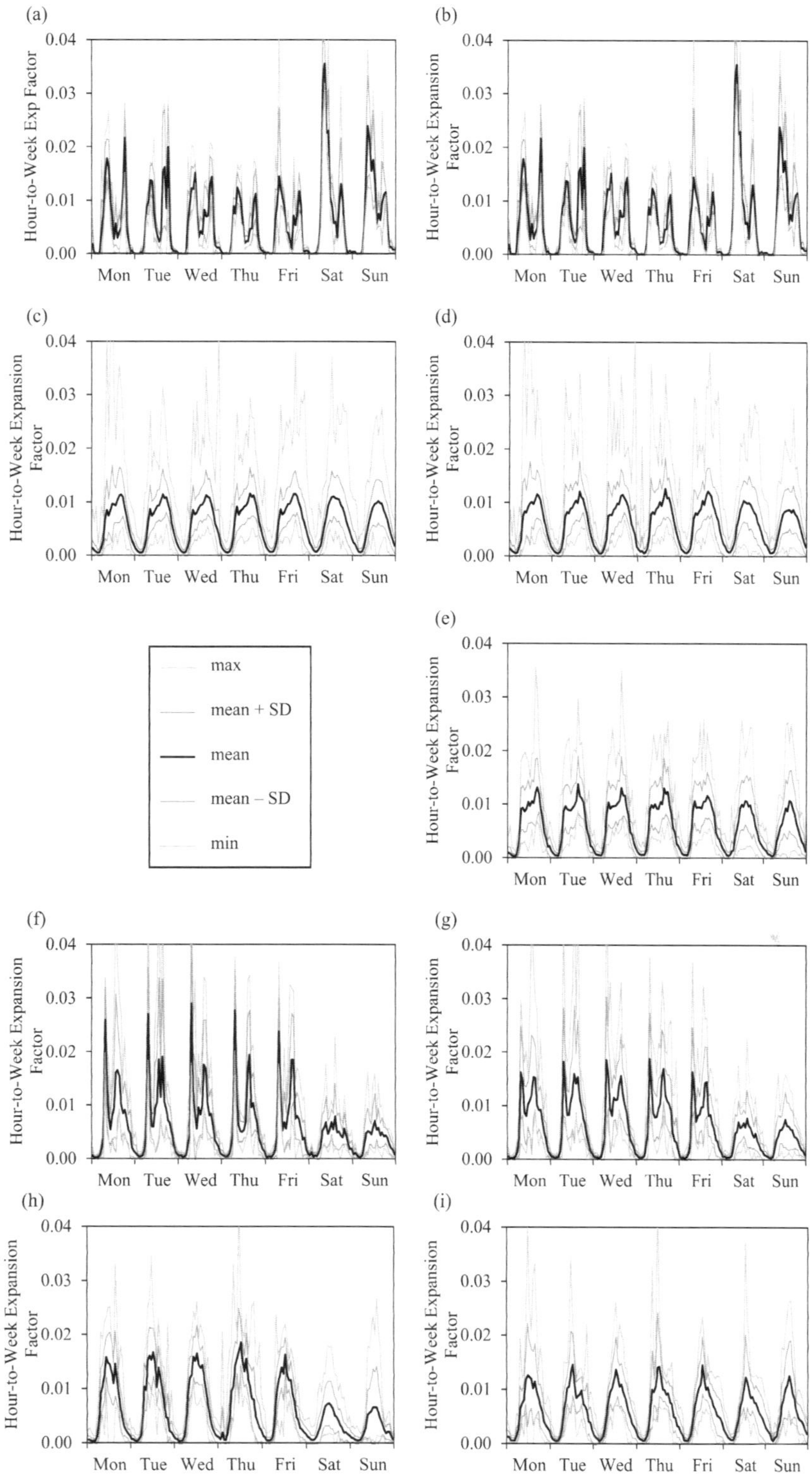

Figure 4. Mean and distribution of activity patterns by empirical cluster (left) and land use group (right): (a) Empirical Cluster 1 (6 sites) (trail-like trend); (b) Trail Land Use (6 sites); (c) Empirical Cluster 2 (73 sites) (commercial-like trend); (d) Commercial Land Use (49 sites); (e) Other Land Use (23 sites); (f) Empirical Cluster 3 (8 sites) (school-like trend); (g) School Land Use (12 sites); (h) Empirical Cluster 4 (16 sites) (CBD-like trend); (i) CBD Land Use (13 sites).

Table 3. Cross-Tabulation of Count Sites by Land Use and Empirical Clusters

Empirical clusters (EC4)	Land use groups: Commercial (49 locs)	Other (23 locs)	CBD (13 locs)	School (12 locs)	Trail (6 locs)	All locations (103 locs)
1	0%	0%	0%	0%	100%	6%
2	86%	83%	62%	33%	0%	71%
3	4%	0%	0%	50%	0%	8%
4	10%	17%	38%	17%	0%	16%
Total	100%	100%	100%	100%	100%	100%

Note: EC4 = best fit empirical clusters (4 clusters); locs = locations.

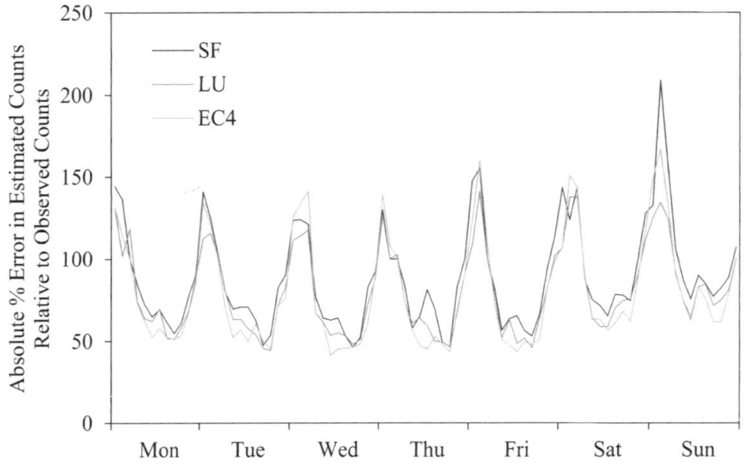

Figure 5. Expansion accuracy by hour of week for 2-h counts.

$$\epsilon^{c}_{i,t,d} = \frac{\left|\frac{\bar{v}_{i,t,d}}{\gamma^{c}_{i,t,d}} - \sum_{t=0}^{23}\sum_{d=0}^{6}\bar{v}_{i,t,d}\right|}{\sum_{t=0}^{23}\sum_{d=0}^{6}\bar{v}_{i,t,d}} * 100 = \left|\frac{\bar{v}_{i,t,d}}{\gamma^{c}_{i,t,d}} - 1\right| * 100 \quad (4)$$

where

c is the cluster type (SF, LU, or EC4) that is being evaluated

i is the location whose expansion by cluster type c is being evaluated relative to the average, observed, location-specific trend

t is the relevant 1-h time period that is being used for estimating the weekly count

d is the corresponding day of week for which a 1-h count may be available

$\sum_{t=0}^{23}\sum_{d=0}^{6}\bar{v}_{i,t,d}$ is the average observed weekly count

Figure 5 indicates the variation in the expansion accuracy by hour of week by plotting the 90th percentile of estimated counts as a percentage of observed counts across all locations. The results reveal that the accuracy, irrespective of the choice of clustering approach, varies significantly by time of day. For instance, expansion based on 2-h night-time counts leads to an estimated count that may have an error of over 150% of the observed count. In contrast, for a daytime count (e.g., noon to 2 p.m.), this error drops to less than 60%. As daytime periods account for a larger percentage of the weekly count totals, a small change in pedestrians counted during a night-time period would have a larger impact on the estimated weekly count than a similar small change in pedestrians counted during the day. Additionally, the comparisons of the clustering approaches reveal that the empirical clustering approach (EC4) may have a marginally lower error for expanding daytime counts.

Similarly, as the duration of the short-term count increases, the variation in the estimated counts generally decreases, as the fluctuations corresponding to individual hours begin to get absorbed into potentially more robust

Table 4. Comparison of Absolute Percentage Error by Short-Term Count Duration and Expansion Factor Approach

	Median absolute % error			90th percentile absolute % error		
Short-term duration	SF	LU	EC4	SF	LU	EC4
Evening peak (4 to 6 p.m.)	25.1%	24.4%	22.7%	71.4%	65.7%	67.2%
Morning and evening peak (7 to 9 a.m. and 4 to 6 p.m.)	23.6%	20.9%	21.2%	56.4%	52.9%	52.4%
Daytime (7 a.m. to 7 p.m.)	13.0%	11.3%	9.8%	35.7%	31.7%	27.6%

Note: SF = single factor; LU = land use groups; EC4 = best fit empirical clusters (4 clusters).

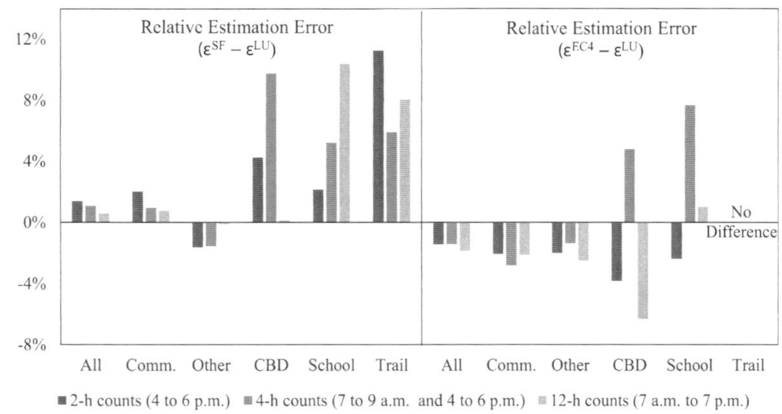

Figure 6. Median absolute difference in percent estimation error relative to land use.

multi-factor expansion factors. For the subsequent analysis, we consider the following durations, based on common count periods:

- Evening peak (4 to 6 p.m.)
- Morning and evening peak (7 to 9 a.m. and 4 to 6 p.m.)
- Daytime (7 a.m. to 7 p.m.)

Short-Term Period Expansion Accuracy

Table 4 presents a comparison of the performance of the clustering approaches by short-term count period. Here, the results indicate that the EC4 has a relatively lower median and 90th percentile absolute percent error among the three approaches for the evening peak and daytime peak expansion, but LU has a slightly lower error on the morning and evening peak. However, as the short-term count durations are increased, 90th percentile estimates reduce dramatically across all clustering approaches. Several other studies have also found greater accuracy from longer-duration counts (*4, 6, 16*). The relatively better performance of the EC approach may also be a case of overfitting, considering that these are in-sample tests.

Although the median absolute error results in Table 4 might indicate a small benefit to clustering expansion factors, they do not account for the difference in performance according to cluster. To examine this variation, we calculated the median absolute estimation error relative to LU for each count location. For example, the left side of Figure 6 shows the difference between the percent estimation errors for the LU and SF approaches ($\epsilon^{SF} - \epsilon^{LU}$). Bars above zero indicate clusters in which the LU clusters perform relatively better. The overall difference shown in the first set of bars (All) is under 2%, but the CBD, School, and Trail groups have median errors as much as 10% less than the SF approach. In particular, the estimates for the Trail group are consistently better with the LU than the SF. This difference is likely due to the greater levels of weekend activity than other sites, which would mean that an SF approach would likely underestimate total activity. The Other group performs slightly better with the SF approach. Comparing the LU and EC approaches on the right side of Figure 6, we see that the empirical clusters do not consistently outperform the LU groups. The School and CBD groups perform better, especially when using peak-period counts.

Considerations

We used in-sample analysis to judge the performance of the different clustering approaches in order to maximize the amount of data used for the expansion factors, but this approach leads to lower error than would be seen in actual practice. Testing out-of-sample count locations also may have been less favorable to the EC approach, due to possible overfitting. It would, however, have presented an additional challenge to assign out-of-sample locations to specific empirical clusters. Future research should explore using of out-of-sample count locations.

The passive infrared counting devices used for this study cannot distinguish between pedestrians and bicyclists. For the trail locations, in particular, this shortcoming may have significantly altered the daily activity patterns captured because of the different trends in bicyclist activity on those facilities.

Due to the varied nature of the built environment, it is not practical to capture or understand all the attributes of a site that will affect the activity patterns. Therefore, we expect to encounter some error in any weekly estimates we create from expansion factors. Some additional influences on pedestrian and bicycle activity patterns have been quantified in previous studies, such as weather and other seasonal factors (e.g., university campus sessions and breaks) (*1, 6, 17–19*). Other influences on pedestrian and bicycle activity, such as festivals, athletic events, and local economic changes, are very difficult to capture in expansion factors.

The clustering approach used in this study assumes that each location's pedestrian activity is influenced by a dominant LU characteristic. However, such an assumption may be harder to apply to intersections that have a mixture of LUs occurring within the vicinity. Thus, future extensions of this work could look into how a location's diverse LU characteristics can be translated into appropriate mixed-use expansion factors.

The long-term count locations used in this study comprise a convenience sample and thus are not necessarily representative of the types of sites present in any particular city. This study provides guidance for LU classification and EC methods that other communities can follow to develop expansion factors.

Finally, the error analysis does not consider the variation within the average location-specific trends.

Conclusion

Expansion factors based on the trends in long-term count data are a useful tool to estimate daily, weekly, or annual volumes from short-term counts, but SFs do not differentiate locations by activity pattern. This paper compares two approaches to grouping expansion factors for the creation of weekly pedestrian volume estimates. The LU classification approach is intuitive and easy to apply to the short-term count sites, but it produced greater error in the weekly volume estimates. The EC approach of using k-means clustering to group locations with similar activity patterns, though methodologically sound, presents a challenge for determining which short-term count locations to apply each group to.

Our results provide useful guidance for practitioners on creating factor groups. Both the LU and EC approaches provided better weekly pedestrian volume estimates than the SF approach of taking the average of all locations. Therefore, using more than one factor group should produce more accurate results when there is variation in the activity patterns among long-term count locations. Further, the differences between the estimation errors for the LU and EC approaches were modest, so it may be beneficial to use the intuitive and practical approach to classify pedestrian activity patterns by surrounding LUs. LU groupings can be refined with insights from the EC results, thus creating improved estimates while maintaining the ease of application.

Comparison of the estimation errors for expansion over typical short-term count periods also provided insights on when counts should be collected. Peak-period counts generally produced weekly volume estimates with lower error than off-peak counts. Thus, ideal times for short-term counts are during peak activity periods. Weekly volume estimates based on longer-duration counts (e.g., 12 h) are generally more accurate than weekly volume estimates based on shorter-duration counts (e.g., 2 h). There was a roughly 50% reduction in median and 90th percentile error estimates between the morning and evening peak and daytime counts.

Future studies should investigate the use of quantitative measures of land use from spatial data for the LU classification. Although visual identification of LU for a limited number of long-term count locations is practical, the task applied to a larger number of short-term count locations would be time consuming and may invite data collector error.

Acknowledgments

This research was funded by the California Department of Transportation. The authors thank Louis Huang of UC Berkeley SafeTREC for his help processing the count data used in this study. We are grateful to Rachel Carpenter of the California Department of Transportation for her assistance in obtaining the data used in this study, and to the Alameda County Transportation Commission, the Fresno Council of Governments, the Los Angeles Department of Public Health, and the San Diego Association of Governments for sharing their automated count data.

Author Contributions

The authors confirm contribution to the paper as follows: study conception and design: JG, AM, RS, OG; data collection: JG; analysis and interpretation of results: JG, AM; draft manuscript preparation: JG, AM, RS. All authors reviewed the results and approved the final version of the manuscript.

References

1. Schneider, R. J., L. S. Arnold, and D. R. Ragland. A Methodology for Counting Pedestrians at Intersections: Using Automated Counters to Extrapolate Weekly Volumes from Short Manual Counts. *Transportation Research Record: Journal of the Transportation Research Board*, 2009. 2140: 1–12.
2. Schneider, R. J., T. Henry, M. F. Mitman, L. Stonehill, and J. Koehler. Development and Application of the San Francisco Pedestrian Intersection Volume Model. *Transportation Research Record: Journal of the Transportation Research Board*, 2012. 2299: 65–78.
3. Nordback, K., W. E. Marshall, B. N. Janson, and E. Stolz. Estimating Annual Average Daily Bicyclists: Error and Accuracy. *Transportation Research Record: Journal of the Transportation Research Board*, 2013. 2339: 90–97.
4. Nordback, K., W. E. Marshall, and B. N. Janson. *Development of Estimation Methodology for Bicycle and Pedestrian Volumes Based on Existing Counts*. Colorado Department of Transportation, Denver, CO, 2013. http://www.coloradodot.info/programs/research/pdfs/2013/bikecounts.pdf/view. Accessed October 2013.
5. Hankey, S., G. Lindsey, and J. Marshall. Day-of-Year Scaling Factors and Design Considerations for Nonmotorized Traffic Monitoring Programs. *Transportation Research Record: Journal of the Transportation Research Board*, 2014. 2468: 64–73.
6. Nosal, T., L. F. Miranda-Moreno, and Z. Krstulic. Incorporating Weather: A Comparative Analysis of Average Annual Daily Bicyclist Estimation Methods. *Transportation Research Record: Journal of the Transportation Research Board*, 2014. 2468: 100–110.
7. Ryus, P., E. Ferguson, K. M. Laustsen, R. J. Schneider, F. R. Proulx, T. Hull, and L. Miranda-Moreno. *NCHRP Report 797: Guidebook on Pedestrian and Bicycle Volume Data Collection*. Transportation Research Board of the National Academies, Washington, D.C., 2014.
8. Liggett, R., H. Huff, R. Taylor-Gratzer, N. Wong, D. Benitez, T. Douglas, J. Howe, J. Cooper, J. Griswold, D. Amos, and F. Proulx. *Bicycle Crash Risk: How Does It Vary, and Why?* Caltrans Task No. 2801. California Department of Transportation, Sacramento, CA, 2016. http://www.lewis.ucla.edu/wp-content/uploads/sites/2/2016/08/Final-Report-to-Caltrans-Bicycle-Crash-v3.pdf. Accessed July 2017.
9. Federal Highway Administration. *Traffic Monitoring Guide*. 2013. http://www.fhwa.dot.gov/policyinformation/tmguide/tmg_fhwa_pl_13_015.pdf. Accessed September 2013.
10. Hocherman, I., A. S. Hakkert, and J. Bar-Ziv. Estimating the Daily Volume of Crossing Pedestrians from Short Counts. *Transportation Research Record: Journal of the Transportation Research Board*, 1988. 1168: 31–38.
11. Ryan, S., D. E. Sidelinger, S. Saitowitz, D. Browner, S. Vance, and L. McDermid. Designing and Implementing a Regional Active Transportation Monitoring Program through a County-MPO-University Collaboration. *American Journal of Health Promotion*, Vol. 28, 2014, S104–S111.
12. Nordback, K., and M. Sellinger. *Methods for Estimating Bicycling and Walking in Washington State*. WA-RD 828.1. Prepared for Washington State Department of Transportation, Olympia, WA, 2014. http://www.wsdot.wa.gov/research/reports/fullreports/828.1.pdf. Accessed May, 2014.
13. Miranda-Moreno, L. F., T. Nosal, R. J. Schneider, and F. Proulx. Classification of Bicycle Traffic Patterns in Five North American Cities. *Transportation Research Record: Journal of the Transportation Research Board*, 2013. 2339: 68–79.
14. Genolini, C., and B. Falissard. KmL: k-Means for Longitudinal Data. *Computational Statistics*, Vol. 25, No. 2, 2010, pp. 317–328.
15. Caliński, T., and J. Harabasz. A Dendrite Method for Cluster Analysis. *Communications in Statistics-Theory and Methods*, Vol. 3, No. 1, 1974, pp. 1–27.
16. Milligan, C., R. Poapst, and J. Montufar. Performance Measures and Input Uncertainty for Pedestrian Crossing Exposure Estimates. *Accident Analysis and Prevention*, Vol. 50, 2013, pp. 490–498.
17. Aultman-Hall, L., D. Lane, and R. R. Lambert. Assessing Impact of Weather and Season on Pedestrian Traffic Volumes. *Transportation Research Record: Journal of the Transportation Research Board*, 2009. 2140: 35–43.
18. Miranda-Moreno, L. F., and A. C. Lahti. Temporal Trends and the Effect of Weather on Pedestrian Volumes: A Case Study of Montreal, Canada. *Transportation Research Part D: Transport and Environment*, Vol. 22, 2013, pp. 54–59.
19. Sears, J., B. S. Flynn, L. Aultman-Hall, and G. S. Dana. To Bike or Not to Bike: Seasonal Factors for Bicycle Commuting. *Transportation Research Record: Journal of the Transportation Research Board*, 2012. 2314: 105–111.

The Standing Committee on Highway Traffic Monitoring (ABJ35) peer-reviewed this paper (18-05579).

Article

TRR
JOURNAL OF THE TRANSPORTATION RESEARCH BOARD

Challenges in Monitoring Regional Trail Traffic

Transportation Research Record
2018, Vol. 2672(43) 98–109

DOI: 10.1177/0361198118787996
journals.sagepub.com/home/trr
SAGE

Greg Lindsey[1], Lila Singer-Berk[1], Jeffrey S. Wilson[2], Eric Oberg[3], and Tracy Hadden-Loh[4]

Abstract

This study reports traffic monitoring results at 30 locations on a 972-mi shared-use trail network across the east-central United States. We illustrate challenges in adapting the principles in the Federal Highway Administration's *Traffic Monitoring Guide* to a regional trail network. We make four contributions: 1) we use factor analysis and k-means clustering to implement a stratified random process for selecting monitoring sites; 2) we illustrate quality assurance procedures and the challenges of obtaining valid results from a multi-state monitoring system; 3) we describe variation in trail traffic volumes across five land use classes in response to daily weather and seasons; and 4) we report two performance measures for the network: annual average daily trail traffic and trail miles traveled. The Rails to Trails Conservancy deployed passive infrared traffic monitors in 2015 through 2017. Site-specific regression models were used to impute missing daily traffic volumes. The effects of weather were consistent across land use classes but the effects of temporal variables, including weekend and season of year, varied. A plan for short-duration monitoring is presented. Results confirm the FHWA monitoring principles and the difficulties of implementing them regionally.

Transportation planners need information about demand for multi-use trails to plan new facilities, assess safety of facilities, prioritize investments in projects, and operate and maintain existing facilities. Despite progress, assessments of comprehensive monitoring programs remain rare, and managers need examples of approaches to designing monitoring networks, data collection and management, and interpretation.

This study reports the results of traffic monitoring at 30 locations between 2015 and 2017 on a 972-mi multi-use trail network in Ohio, Pennsylvania, New York, and West Virginia by the Rails to Trails Conservancy (RTC) and the Industrial Heartland Trails Coalition (IHTC). Our objective is to illustrate challenges in adapting the Federal Highway Administration's *Traffic Monitoring Guide* (*TMG*) to a regional trail network (*1*).

Challenges in Nonmotorized Traffic Monitoring: Literature Review

In 2013, in response to pioneering efforts by engineers and advocates, the FHWA added a chapter on nonmotorized traffic monitoring to the *TMG*, the guidance document that establishes national monitoring principles. Chapter 4 "Traffic Monitoring for Nonmotorized Traffic" builds on the principles of monitoring motorized traffic but notes nonmotorized traffic is more variable and requires different procedures (*1*). The basic approach, however, is analogous:

- Establish monitoring objectives
- Determine modes of traffic to be monitored
- Select monitoring sites, including permanent and short-duration stations
- Determine the type(s) of devices to be deployed
- Implement monitoring following recommended guidelines
- Follow recommended analytic procedures to ensure validity of data
- Use factors derived from permanent monitoring stations to extrapolate short-duration counts and estimate annual average daily bicyclists (AADB), pedestrians (AADP) or mixed-mode, undifferentiated nonmotorized traffic

The *TMG* also lists steps in implementation of permanent and short-duration monitoring programs. This approach presents new analytic challenges and is the focus of research.

[1]Humphrey School of Public Affairs, University of Minnesota, Minneapolis, MN
[2]Department of Geography, Indiana University-Purdue University Indianapolis, Indianapolis, IN
[3]Rails-to-Trails Conservancy, Yellow Springs, OH
[4]Center for Real Estate and Urban Analysis, George Washington University, Washington, D.C.

Corresponding Author:
Address correspondence to Lila Singer-Berk: singe175@umn.edu

This review summarizes research that informs choices in program implementation and advances procedures used to analyze counts.

Program Design and Evaluation

Several recent publications inform monitoring implementation. Griffin et al. (*2*) provide an overview of considerations in nonmotorized traffic monitoring. The Institute for Transportation Research and Education (ITRE) at North Carolina State University, in collaboration with the North Carolina Department of Transportation, is leading one of the largest state-level bicycle and pedestrian data collection programs in the United States (*3*, *4*). Minge et al. recently published a guide to bicycle and pedestrian data collection for the Minnesota Department of Transportation (MnDOT) (*5*). Progress in institutionalizing nonmotorized traffic monitoring in Colorado, Minnesota, and Oregon has been described (*6*), and other states, including Washington and Vermont have initiated programs. Much of the implementation is by regional and local governments, such as San Francisco and New York City, and is described on websites and in the professional literature. The Delaware Valley Regional Planning Commission (DVRPC) manages one of the largest regional monitoring programs in the United States and makes estimates of AADB and AADP available through its website (*7*). The Mid-Ohio Regional Plan Commission (MORPC) maintains a network of monitors on the regional trail system and has reported estimates of annual average daily trail traffic (AADTT) and trail miles traveled for the network (*8*, *9*). The City of Vancouver, British Columbia has implemented monitoring and is reporting bicycle traffic volumes (*10*).

Monitoring Site Selection

The *TMG* specifies three key decisions: "differentiating between pedestrian and bicycle traffic," "selecting representative permanent count locations," and "selecting optimal installation locations" (*1*). With respect to selecting locations, the *TMG* cautions against selecting sites with the heaviest volumes but, other than recommending a review of existing count data, does not specify how to determine representative locations. Agencies have approached this challenge differently. In North Carolina, researchers completed a study to determine the number and location of counters to obtain valid, reliable estimates of AADB. Because the cost exceeded available funds, ITRE and NCDOT established the objective of developing permanent stations at locations in urban, rural, and near-university areas believed to have commuting, recreation, and mixed traffic patterns (*3*, *4*). ITRE inspects potential locations and conducts field tests before permanent counters are installed. Their goal is to implement a dozen or more counters per region. MnDOT incorporated existing counters, and established at least one bicycle monitor on streets and one mixed-mode monitor on trails in each administrative region. These sites serve as index sites to illustrate trends and demonstrate how counts can inform planning and engineering (*11*). Sites were selected with local partners to ensure results had practical significance.

Each of these approaches has involved stratified systematic or purposeful selection of locations, or both stratified systematic and purposeful selection, absent measures of actual representativeness for the networks of interest. Davis et al. demonstrated the feasibility of a stratified, randomized site selection for purposes of estimating regional bicycle miles traveled, but their approach was not designed to be augmented with short-duration counts to characterize network flows (*12*). Randomization of site selection typically has not been implemented by public agencies.

Monitoring Equipment Selection

A National Cooperative Highway Research Program report, "Methods and Technologies for Pedestrian and Bicycle Volume Data Collection," has become the authoritative guide to monitoring technologies (*13*, *14*). This report has been augmented by research papers. Nordback et al. and Brosnan et al. have evaluated the performance of inductive loops and pneumatic tubes, respectively (*15*, *16*). In general, tradeoffs among types of technologies are understood – they involve the need for mode-specific information, accuracy, cost, labor for data collection, capacity for remote reporting, and vendor support. Trends in deployment are becoming clear: inductive loops are being used to count bicycles at permanent installations; pneumatic tubes are used for short-duration bicycle counts; infrared sensors are used to count pedestrians and trail users, sometimes with inductive loops. Automated video processing remains the Holy Grail of monitoring, but it has not been implemented widely.

Data Quality Management

Questions in data quality management include validation of counters, whether to correct for systematic error associated with sensors (e.g., undercounts due to occlusion), how to implement quality assurance/quality control (QAQC) procedures, and whether to impute missing counts. Analysts have addressed these questions in different ways, and standard procedures are being developed.

All vendors recommend in-field validation of equipment following installation, but the duration of validation and periods for revalidation vary. Personnel deploying integrated inductive loop and infrared sensors at permanent stations may observe traffic for 1 to 2 h following installation, whereas personnel deploying portable equipment may validate less than an hour. Some researchers have adjusted all hourly counts to correct for occlusion (*17–19*). Judgment must be exercised in applying correction factors, however, to avoid changing counts when inappropriate (e.g., for zero counts).

The *TMG* describes the importance of applying QAQC procedures to nonmotorized counts, but notes that new checks need to be developed because of the variability of nonmotorized counts (*1*). Development of QAQC protocols for nonmotorized counts is an active area of research. Turner and Lasley recommend, at minimum:

- Visual inspection of data
- Use of pre-specified criteria to identify potential outliers
- Assessment of zero counts
- Use of professional judgment to censor counts believed to be invalid (*20*)

Based on their North Carolina experience, Jackson et al. recommend several protocols: weekly visual inspections to ensure prompt identification of problems, development of hourly data checks, interquartile checks to identify outliers, and automated procedures for flagging suspect data (*4*). Minge et al. illustrate application of QAQC procedures on data from 12 sensors in Minnesota and the effects of censoring counts on estimates of annual average daily traffic (AADT) (*5*). They show that many outliers may be valid counts associated with events and that it is difficult to differentiate valid and invalid hourly zero counts, particularly in winter. For low-volume sites (e.g., less than 100 counts per day), the cost of implementing checks must be weighed against the practical significance of changes in estimates of volumes that might result from application of checks.

One approach to missing observations is to ignore them and estimate the statistics of interest with available data. Another approach is to estimate values for missing days from existing data. This can be done by assuming the volumes are the same as the volumes for closely related days with valid counts or by more sophisticated statistical methods. For example, a missing count for a Thursday in March could be estimated as the mean of existing counts for Thursdays in March. This approach controls for day of week and season but not weather. An alternative approach is to estimate a regression equation that controls for day of week and weather, obtain weather data for the days with missing counts, and use the equations to estimate missing values (*9, 17, 19*). This procedure involves the judgment that errors associated with imputing are preferable to errors caused by the missing values.

Determination of Pattern Groups

The *TMG* observes that nonmotorized traffic can be classified as commuter, recreational, or mixed (i.e., multipurpose) traffic based on day-of-week and hourly relationships and that it is important to take traffic patterns into account when extrapolating short-term counts. Miranda-Moreno et al. illustrate how weekend–weekday and a.m.–noon hour traffic ratios can be used to classify traffic as (1) utilitarian, (2) mixed-utilitarian, (3) mixed-recreational, and (4) recreational (*21*). Analysts in Colorado established a separate factor group for cycling in mountain regions because patterns associated with seasons diverge from those in urban areas. In general, most programs focus on three factor groups based on some variation of the ratios illustrated by Miranda-Moreno (e.g., North Carolina [*3, 4*]; Minnesota [*5*]). Nordback et al. recommend a minimum of five permanent monitors per factor group to minimize error in extrapolation (*22*); data are not available to assess how many programs are consistent with this recommendation.

Computation of Adjustment Factors

The *TMG* illustrates procedures analogous to those used in motorized traffic monitoring for computation of adjustment factors used in extrapolation of short-duration counts (i.e., hour-of-day, day-of-week, and month-of-year factors). Nordback et al. described error in estimating AADB associated with the length of the short-duration counts and the season when short-duration counts were taken (*22*). They recommended counts of at least one week between April and October. Hankey et al. and Nosal simultaneously developed an alternative approach called day-of-year factoring that reduced error in extrapolation compared to the standard approach (*17, 23*). El Esaway subsequently demonstrated the day-of-year factoring approach is preferable because it better captures the effects of weather (*24*). Jessberger et al. recently published an alternative to the standard approach that holds promise (*25*).

Estimation of Performance Indicators

A purpose of the *TMG* is to guide implementation that produces two performance indicators: AADT and annual miles traveled. State (e.g., North Carolina, Colorado, Minnesota), regional (e.g., DVRPC, MORPC), and local (e.g., Vancouver, BC) agencies now are routinely reporting AADB and AADP for permanent monitoring locations, but fewer have implemented the short-duration counts required to characterize networks. Estimates of AADT have been reported for urban trails in Minneapolis, MN, Columbus, OH, and Chicago, IL (*9, 26*).

Implications for Practice

This review shows that principles of motorized traffic monitoring are being applied to nonmotorized traffic monitoring, that state, regional, and local agencies are implementing monitoring programs, but that additional research is needed to improve program quality. Tradeoffs among objectives in monitoring are inevitable. Because most monitoring programs are new, a gap in the literature is the absence of evaluations of programmatic initiatives. Additional studies of monitoring programs can inform future initiatives.

The Industrial Heartland Trail Monitoring Initiative: Approach, Methods, and Results

The IHTC is a coalition of more than 100 public and private, nonprofit organizations supported by the RTC, the National Park Service, and the Pennsylvania Environmental Council in work to establish a 1,400-mi trail network across 48 counties in Pennsylvania, Ohio, New York, and West Virginia. Nearly 1,000 mi of trail already exist. In 2015, RTC initiated a regional monitoring program to support the initiative.

Program Design and Evaluation

RTC established three objectives for the program:

- Document use on existing trails using procedures consistent with *TMG* principles
- Inform comprehensive regional monitoring efforts
- Develop tools to support trail planning, including factors for extrapolating short-duration counts and estimates of network use

RTC consulted with partners, obtained a grant, and consulted with researchers about program design.

Monitoring Site Selection

Considerations in monitoring site selection included geographic variation of land uses adjacent to trails and the desire to monitor sites characteristic of different factor groups (*27*). We chose a stratified random sampling approach to avoid bias toward high volume locations, increase the likelihood of identifying different patterns, and provide a stronger statistical foundation for generalizing results. A land use typology was used because research has shown that both traffic volumes and patterns may be associated with land use.

The final classifications were based on factor analysis scores derived from 16 contextual measures that were classified using a k-means clustering approach. The contextual measures included census data such as population and job density and remote sensing data. For logistical reasons, sample sites were selected from a subset of the entire 605-mi network. We randomly selected six locations within each class for a total of 30 monitoring sites (Figure 1). The number of sites was determined in light of budget constraints and the *TMG* recommendation that factor groups have a minimum of five sites to develop adjustment factors. RTC reviewed the locations to verify feasibility of access and counter installation.

Monitoring Equipment Selection

We selected lower-cost passive infrared monitors because this choice maximized monitoring locations and provided opportunities to engage local partners. The sensors can be adapted for future short-duration monitoring as portable counters. The rationale for choosing mixed-mode over mode-specific sensors was that it was more important to gain understanding of total traffic at more locations than mode-specific traffic at fewer locations. Although an integrated inductive loop-passive infrared system was installed at one location, mixed-mode counts are reported here.

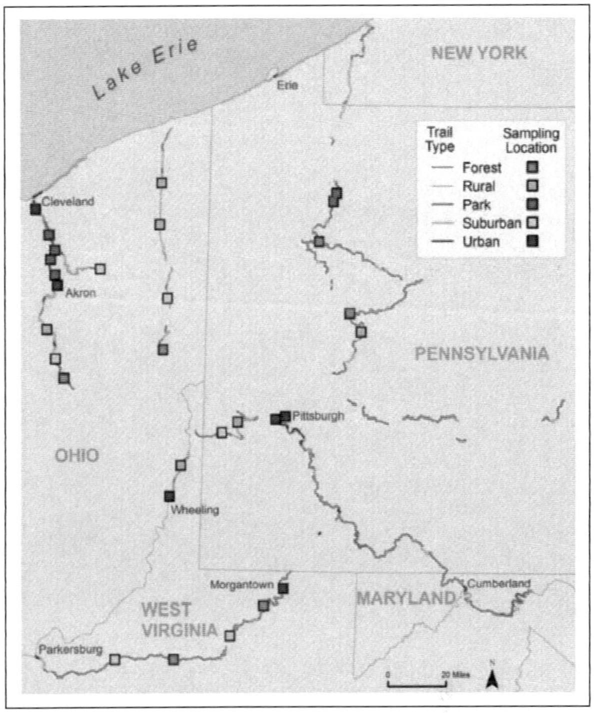

Figure 1. Monitoring locations in the IHTC trail network.

Data Quality Management

Challenges in data collection and management included deployment, in-field validation of counter operations, application of QAQC procedures, and management of missing observations. Although RTC confirmed monitors were operating following deployment and periodically checked monitors for specific reasons, most monitors were not systematically analyzed until the end of the monitoring period because project partners had limited time. When researchers reviewed the data in the winter of 2017, it was the first comprehensive review. Although the objective was to obtain at least 1 year of data at each of the 30 locations, only 22 monitors (73%) were deployed for a full-year, mainly because of logistics and delays in deployment, including coordination with partners (Table 1). The days of counts for the sites with less than 1 year of deployment ranged from 116 to 364.

We followed recommended QAQC procedures: visual inspection, application of a heuristic to identify potential

Table 1. Summary Data for Counts at 30 Locations on IHTC Trail Network

Counter	Class	Trail/location	Days deployed	Days with counts (pre-QAQC)	Days with valid counts (post-QAQC)	Days with valid zero counts (pre-QAQC)	Maximum valid daily count	Minimum valid daily count	ADT: days with counts (pre-QAQC)	ADT: days with valid counts (post-QAQC)	% change in ADT after QA/QC
1	Urban	Ohio and Erie Canal Towpath (OH)	346	257	257	0	724	26	235	235	0%
7	Urban	Ohio and Erie Canal Towpath (OH)	443	443	443	0	801	3	101	101	0%
15	Urban	Wheeling Heritage Trail (WV)	364	328	328	0	842	1	244	244	0%
16	Urban	Three Rivers Heritage Trail (PA)	241	241	235	0	2,901	8	402	268	−50%
17	Urban	Three Rivers Heritage Trail (PA)	116	116	116	0	650	4	123	123	0%
25	Urban	Mon River Trail (WV)	369	369	369	0	905	8	324	324	0%
5	Suburban	Portage Hike and Bike Trail (OH)	392	392	225	0	391	2	820	138	−493%
9	Suburban	Ohio and Erie Canal Towpath (OH)	444	442	362	2	928	0	6,284	81	−7,628%
12	Suburban	Mill Creek Metro Parks Bikeway (OH)	458	458	458	0	826	1	189	189	0%
19	Suburban	Panhandle Trail (PA)	353	351	351	0	461	2	54	54	0%
27	Suburban	West Fork River Trail (WV)	442	442	442	22	276	0	31	31	0%
29	Suburban	North Bend Rail-Trail (WV)	442	442	442	15	330	0	28	28	0%
8	Rural	Ohio and Erie Canal Towpath (OH)	444	444	444	3	1,376	0	140	140	0%
10	Rural	Western Reserve Greenway (OH)	393	393	393	0	297	1	74	74	0%
11	Rural	Western Reserve Greenway (OH)	401	359	359	1	522	0	154	154	0%
14	Rural	Brooke Pioneer Trail (WV)	364	364	364	10	172	0	35	35	0%
18	Rural	Montour Trail (PA)	353	352	352	11	561	0	60	60	0%
20	Rural	Armstrong Trail (PA)	353	353	353	34	256	0	20	20	0%
2	Parks	Ohio and Erie Canal Towpath (OH)	368	368	368	0	1,137	5	254	254	0%
3	Parks	Ohio and Erie Canal Towpath (OH)	368	368	368	0	2,988	10	596	596	0%
4	Parks	Ohio and Erie Canal Towpath (OH)	368	368	368	0	2,466	10	422	422	0%
6	Parks	Ohio and Erie Canal Towpath (OH)	368	368	368	0	731	11	175	175	0%
23	Parks	Oil Creek State Park Trail (PA)	367	367	367	9	452	0	65	65	0%
24	Parks	Oil Creek State Park Trail (PA)	379	378	378	50	326	0	34	34	0%
13	Forest	Little Beavercreek Greenway (OH)	240	240	240	3	534	0	63	63	0%
21	Forest	Armstrong Trail (PA)	241	241	241	29	99	0	15	15	0%
22	Forest	Allegheny River Trail (PA)	382	379	379	33	3,064	0	94	94	0%
26	Forest	Mon River Trail (WV)	442	442	442	55	435	0	21	21	0%
28	Forest	North Bend Rail-Trail (WV)	442	442	442	171	93	0	4	4	0%
30	Forest	Ohio and Erie Canal Towpath (OH)	444	444	444	1	323	0	54	54	0%

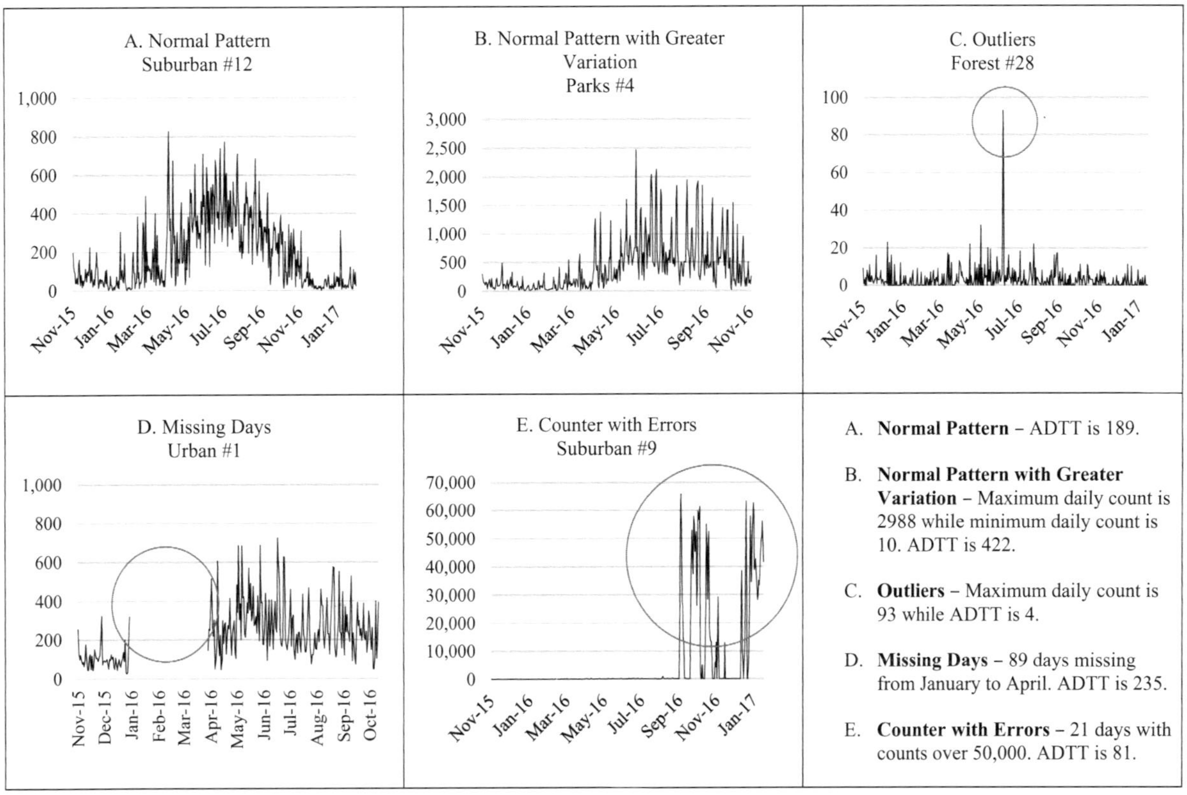

Figure 2. Examples of daily counts at five locations prior to QAQC.

outliers, evaluation of counts of zero, and use of judgment (*20*). Visual inspection of daily volume graphs identified sites with missing data and patterns that indicated sensor malfunction. Inspection of data (illustrated below) showed seven monitors (23%) had missing days, apparently associated with insect infestation, vandalism, or other malfunction (Table 1). Following additional QAQC checks, outliers were censored at three locations, resulting in 19 monitors (63%) with valid counts for at least 365 days. Across locations, monitors were deployed for 11,127 days; counts were obtained for 10,951 days, 98% of the deployment periods for all counters. The total number of days with counts deemed valid was 10,698 (96% of the days of deployment). RTC decided to report published, approximate values for accuracy for passive infrared sensors rather than develop correction factors for each location.

To illustrate the problems identified during QAQC, we present daily totals for the period of record for five sites (Figure 2). Graphs A and B show "normal" patterns that reflect daily, weekly, and seasonal variation. The records appear complete, and the variation does not appear abnormal. Graph C illustrates a case with a potential outlier: most values for the period of record are below 100 but, for a single day, values increase more than fourfold. Graph D is from a site where no data were recorded for a 4-month period. Graph E has a jump to counts of more than 60,000 per day after a 10-month period of daily counts below 1,000, an obvious error. The team used professional judgment to censor daily counts that were associated with sensor malfunction.

To identify potential outliers, we identified counts equal to or greater than three standard deviations above the mean daily count for the period of record. To decide whether to censor questionable counts, we computed the ADT with and without the potential outliers. For counters with ADT below 90, all recorded counts were retained because, even including potential outliers, the site was low volume. The rationale for the threshold of 90 was that inclusion of the potentially invalid data would have minimal practical implications for ADT. For counters with ADT above 90, the flagged values were retained if the percentage reduction in ADT was less than 15% (an arbitrary threshold). If the reduction in ADT was greater than 15%, we assessed whether favorable weather conditions may have induced unusual use and searched for events that may have occurred. Professional judgment then was applied to retain or censor counts. We inspected days with counts of zero, but did not eliminate any zero count days.

Another challenge in data quality management involved managing the 11 sites with missing observations (which ranged from 1% to 26%). For purposes of estimating

Table 2. Pooled Regression Models of Daily Counts by Land Use Class

	Urban (n = 1,740)	Suburban (n = 2,227)	Rural (n = 2,184)	Parks (n = 2,184)	Forest (n = 2,193)	All sites (n = 10,528)
Dependent variable: ADT	216	87	81	258	42	137
Average dew point (°F)	−0.014***	−0.003	−0.021***	0.001	−0.024***	−0.014***
Average wind speed (knots)	−0.045***	0.068***	0.007	0.070***	−0.018*	0.012***
Maximum temperature (°F)	0.073***	0.069***	−0.010	0.081***	0.077***	0.054**
Maximum temperature squared (°F^2)	−0.0004***	−0.0003***	0.0004***	−0.0005***	−0.0001	−0.0001***
Precipitation (in.)	0.079	−0.587***	−0.349***	−0.795***	−0.361***	−0.444***
Weekend (dummy)	0.165***	0.333***	0.538***	0.672***	0.850***	0.526***
Spring (dummy)	0.552***	0.665***	0.568***	0.648***	0.093	0.458***
Summer (dummy)	0.615***	1.057***	0.512***	1.158***	0.019	0.532***
Fall (dummy)	0.349***	0.705***	0.522***	0.623***	0.020	0.387***
Constant	2.686***	0.280	3.242***	−0.802**	1.573***	1.714***
Pseudo R^2	0.039	0.049	0.037	0.039	0.046	0.028***

Note: Modeling approach: negative binomial regression.
*$p<.01$. **$p<.005$. ***$p<.001$.

AADT, we imputed values for missing days. Research has shown that site-specific regression models including weather and day-of-week variables explain 85% of the variation in daily traffic and can be used to predict traffic with reasonable accuracy (e.g., [*19*]). Using only days with valid counts, we assembled weather data for each day in the period of record and estimated negative binomial weather regression models for each site. Weather data for the missing days were used in site-specific models to predict daily traffic for missing days.

To illustrate model specification and differences in response to weather across land use classes, data in each of five classes were pooled, and six additional models were estimated, one for each class and one with data pooled for all sites (Table 2). The magnitudes of some coefficients vary. For example, weekend daily traffic was statistically significantly higher than weekday daily traffic in every class. Precipitation, which significantly reduced traffic at parks, suburban, rural, and forest sites, had no significant effect on urban traffic volumes, a departure from general patterns.

Determination of Factor Groups

The challenge in this step involved choice of criteria for establishing factor groups. Using days with counts deemed valid (i.e., excluding days with counts estimated with weather models), recommended procedures were adapted to classify sites into three factor groups (*21*). The weekend–weekday average daily traffic ratio and the a.m. (7:00 to 9:00 a.m.) to noon (11:00 a.m. to 1:00 p.m.) average hourly traffic ratio were computed to classify sites. Consistent with the *TMG*, only three factor groups were classified: recreational, commuter, and mixed. Application of these criteria identified only two factor groups: recreation (n = 28) and mixed (n = 2). The sites with mixed traffic patterns included one urban and one parks location.

Computation of Adjustment Factors

We computed hour-of-day, day-of-week, and month-of-year factors that could be applied to estimate AADTT from short-duration counts. Because application of criteria for factor groups revealed little variation in patterns across sites, we computed factors for sites by the land use classification. The monthly average daily trail traffic (MADTT) to AADTT ratios included six sites in each class and the 30 sites pooled together (Figure 3). These graphs were computed using datasets with counts estimated for missing days. Across sites within classes, variability in MADTT/AADTT ratios is greatest with urban and forest classes. The convergence in patterns across classes, however, is evident, with the mean ratios for all classes following a general pattern. The MADTT/AADTT ratios for summer months, which reflect peak volumes and periods when short-duration sampling is most likely to occur, generally are between 1.5 and 2. The mean of means ratio for summertime months for all sites is approximately 1.5, indicating that, on average, daily summertime traffic is 50% higher than AADTT. Hour-of-day graphs for all sites could be used as adjustment factors for short-duration counts (Figure 4).

Estimation of Performance Indicators

Following Davis et al., we multiplied the mean AADTT for each land use classification by the miles of trail within the classification (*12*). This approach, which was used because short-duration counts had not been completed, deviates from standard procedure of multiplying segment counts × segment length. The limitation of this approach is that it does not provide segment-specific estimates of AADTT. Nonetheless, the approach provides an order of magnitude estimate until short-duration counts can be completed. Mean AADTT ranged from 40 at forest locations to 251 and 258 at urban

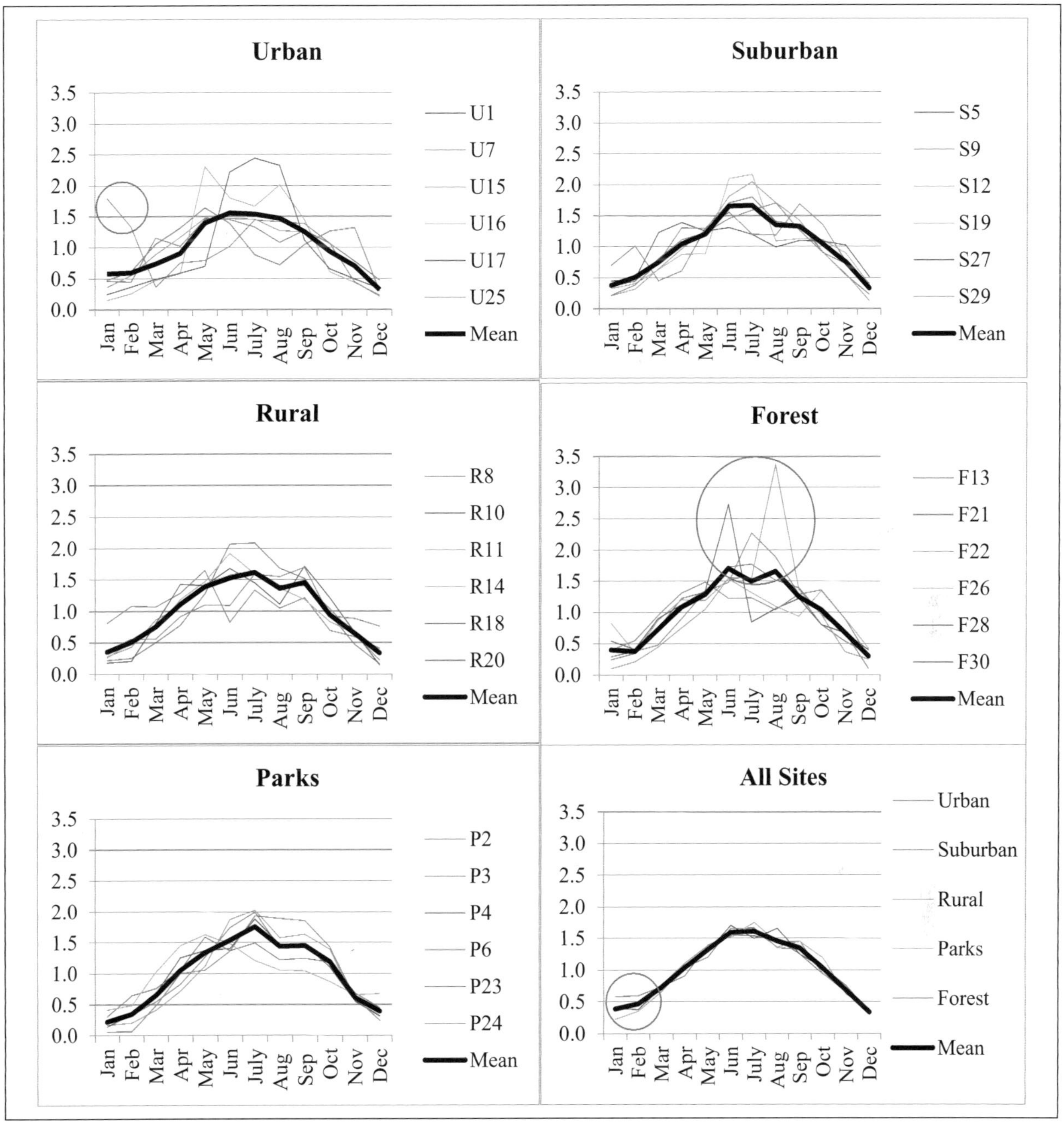

Figure 3. MADTT/AADTT ratios by land use classification.

and parks locations, respectively (Table 3). The estimate of total trail miles traveled in 2016 for the 972-mi network was approximately 29.5 million mi. Urban and park trails, which account for only about 11% of total miles, accounted for 33% of miles traveled. Forest trails, which account for nearly half the network (47%), accounted for less than one-quarter of the trail miles traveled. Sources of uncertainty in this estimate include the error associated in counting (i.e., systematic undercount due to occlusion, the magnitude of which may vary across sites in relation to volume), error associated with imputing missing AADTT values at some sites, and error in extrapolating mean AADTT values to the network.

Permanent and Short-Duration Site Selection

In the future, permanent and new short-duration counts will be used to refine AADTT estimates. RTC plans to retain some of the monitors permanently and to shift others to portable

Figure 4. Hourly weekday and weekend traffic.

Table 3. Performance Indicators for IHTC Trail Network (2016)

				Performance indicators			Measurable results		Proposed permanent counters		Proposed short-duration counters		
Sample class	Number of sample points	Estimated trail miles	AADTT	Estimated trail miles traveled	% of sample points (miles)	% of miles traveled	Rec.	Mixed	Rec.	Mixed	Segment length	Maximum points	Per year
Forest	497	457	40	6,700,000	47%	23%	6	0	3	0	5	157	79
Low intensity and rural	248	228	84	7,000,000	23%	24%	6	0	2	0	2	153	77
Parks	72	66	258	6,200,000	7%	21%	6	0	2	0	2	128	64
Suburban	196	180	90	5,900,000	19%	20%	5	1	3	1	1	36	18
Urban	43	40	251	3,600,000	4%	12%	5	1	2	1	1	65	33
Totals	1056	972	137	29,500,000	100%	100%	28	2	12	2		539	270
							30		14				

Note: Rec. = recreational.

counts to monitor the network. To illustrate tradeoffs in establishing the short-duration monitoring program given a limited fleet of counters, we present an example in which 14 counters are retained as permanent counters and 16 are deployed for short-duration counts. To determine which sites would be retained as permanent monitoring sites, we considered factor group (i.e., recreational, mixed), land use classification, volume, land use, and location within the region, with a goal to include all factor groups and retain three to four monitoring locations in each land use classification. Thus, this example includes both sites with mixed traffic patterns and meets the *TMG* recommendation for recreational patterns but falls short for mixed patterns. With respect to land use, four urban, two suburban, two rural, three park, and three forest monitoring sites were chosen for permanent monitoring. Fewer sites for suburban and rural were selected because their AADTTs were very close (87 and 81, respectively) and their weekend and weekday traffic patterns were similar. A second decision in planning short-duration counts involved determining segments. We assumed it would not be feasible given resource limitations to sample each mile of trail, so the decision was made to vary segment length by land use, which is associated with volume. We assumed segments of 5 mi for forest trails, 2 mi for rural and suburban trails, and 1 mi for urban and park trails, which have the highest volumes and, potentially, greatest variation along consecutive segments. This approach resulted in a maximum of 539 potential sampling points (Table 3). With a May to October monitoring period, 16 counters, and 10 days for each 7-day short-duration count, monitoring of all sites could be completed in two summers (Table 3). If RTC decides to retain more permanent counters, or obtains funding for portable counters, this plan could be revised. This example illustrates the feasibility of completing monitoring in the network and the tradeoffs required to achieve this objective.

Conclusions and Implications for Practice

Management of a monitoring program requires application of engineering principles and use of professional judgment, subject to economic constraints, in diverse geographic and socio-political settings. Managers must make tradeoffs in pursuit of data. Jackson et al. observe that "data-wranglers" are needed to launch and manage nonmotorized traffic monitoring programs (*3*, *4*). Data-wrangling is an apt metaphor for the challenges of monitoring require taking charge, rounding up, herding, and eventually, producing results.

This study illustrates the nature of data-wrangling and some of the tradeoffs required in launching a trail monitoring program on a multi-state network. Consistent with the monitoring objectives, budget constraints, and logistics of installation and maintenance, RTC installed 30 counters, but complications resulted in less than 1 year of data for nearly one-quarter of the monitors. Application of QAQC procedures revealed missing data, erroneous counts likely associated with counter malfunction, and outliers judged invalid and censored. Valid counts for a 1-year period were obtained for only 19 (63%) of the 30 sites. Overall, however, valid counts were obtained for 96% of the days deployed, indicating more days were missed getting the program running than because of problems following deployment.

These results led to analytic choices that involved tradeoffs among sources of error in subsequent estimates. For example, regression models were used to estimate missing counts for 11 sites prior to computing monthly adjustment factors. We could have computed these factors for only the

19 sites with complete, year-long records or for only sites with, for example, 300 or more days of valid counts ($n = 24$). The tradeoff involved whether to keep sites but accept the error associated with imputing missing counts or to reject sites and rely on fewer sites with complete records. The objective of characterizing traffic at as many locations as possible led to the decision to impute missing values.

Each of these judgments has implications for the validity of the performance measures. One approach to assessment is whether the results are consistent with theory and make intuitive sense. As volumes are highest at parks and urban locations, lower in suburban and rural areas, and lowest in more remote forest areas, the results reflect population density gradients and meet this criterion. Most of the network occurs in rural and remote areas, thus, the volumes are substantially lower than AADTTs reported for urban networks measured with essentially the same approach (*9*, *26*). Because the randomized selection of locations avoids bias associated with purposeful selection of sites with high volumes, there is reason to believe these measures are representative.

Several features of this initiative are distinctive and inform practice. First, use of advanced geo-spatial analytics to classify potential sampling points in the network and implement stratified-randomized selection of monitoring locations provided insight into volumes of trail use at sites that likely would be ignored in a more purposeful sampling approach. Estimates of AADTT for the network therefore reflect sections of trail with very low volumes. Second, studies of new or best practices often do not document or quantify problems in ways that facilitate comparison across programs. We report problems in implementation and indicators of the reliability and validity of counts produced for the first year of a permanent monitoring program for a large region. Systematic application of QAQC procedures identified missing and invalid counts that were censored from the dataset. Ultimately, valid counts were obtained for 365 days at only 19 (63%) of the 30 locations. Percentages for initiatives of comparable size are not available, but it may be that the geographic scale of this initiative contributed to this outcome.

Third, we show that trail traffic on this network in the east-central United States varies in relation to land use but that patterns converge. Monthly traffic patterns varied more within urban and forest classes than within parks, suburban, and rural classes, but mean ratios across classes were similar. Use of standard metrics indicated 28 sites were characterized by recreational traffic; no locations reflected commuter traffic. Fourth, these results complement previous estimates of AADTT and miles traveled on urban trail networks and contribute to our understanding of variation in trail traffic across the country. Fifth, we illustrate how counters could be redeployed to complete monitoring of the entire network within a 2-year period.

Overall, this study illustrates that the principles in the *TMG* can be adapted for regional trail monitoring. The study also illustrates that, given resource constraints and practical limitations, challenges will be encountered, and tradeoffs will be required in each step of the monitoring process. Decisions in implementation, from selection of sites to application of QAQC procedures to determination of factor groups, affect estimates and decisions based on those estimates. Additional studies of ways in which data wranglers meet these challenges can help inform practice.

Author Contributions

The authors confirm contribution to the paper as follows: study conception and design: TH-L, GL, JW; data collection: EO; analysis and interpretation of results: GL, LS-B; draft manuscript preparation: GL, LS-B. All authors reviewed the results and approved the final version of the manuscript.

References

1. Federal Highway Administration, U.S. Department of Transportation. *Traffic Monitoring Guide: Chapter 4 Traffic Monitoring for Non-motorized Traffic V1*. Federal Highway Administration, U.S. Department of Transportation, 2013. https://www.fhwa.dot.gov/policyinformation/tmguide/tmg_fhwa_pl_13_015.pdf.
2. Griffin, G., K. Nordback, T. Götschi, E. Stolz, and S. Kothuri. *Transportation Research E-Circular E-C183: Monitoring Bicyclist and Pedestrian Travel and Behavior: Current Research and Practice*. Transportation Research Board of the National Academies, Washington, D.C., 2014.
3. Jackson, K. N., E. Stolz, and C. Cunningham. Nonmotorized site selection methods for continuous and short-duration volume counting. *Transportation Research Record: Journal of the Transportation Research Board*, 2015. 2527: 49–57.
4. Jackson, K. N., S. W. O'Brien, S. E. Searcy, and S. E. Warchol. Quality Assurance and Quality Control Processes for a Large-Scale Bicycle and Pedestrian Volume Data Program. *Transportation Research Record: Journal of the Transportation Research Board*, 2017. 2644: 19–29.
5. Minge, E., C. Falero, G. Lindsey, M. Petesch, and T. Vorvick. Bicycle and Pedestrian Data Collection Manual. MnDOT Report No. MN/RC 2017-03. *Minnesota Department of Transportation*, 2017. http://www.dot.state.mn.us/research/reports/2017/201703.pdf.
6. Lindsey, G., K. Nordback, and M. Figliozzi. Institutionalizing Bicycle and Pedestrian Monitoring Programs in Three States. *Transportation Research Record: Journal of the Transportation Research Board*, 2014. 2443: 134–142.
7. Delaware Valley Regional Planning Commission. *Pedestrian and Bicycle Counts*. 2017. http://www.dvrpc.org/webmaps/pedbikecounts/.
8. Lindsey, G., X. Wu, C. Wu, J. Ciabotti, and R. Edwards. *Technical Memorandum 1. COG Trails: Traffic Monitoring Results*. Mid-Oho Regional Planning Commission, Columbus, Ohio, 2015.
9. Wang, J., S. Hankey, X. Wu, and G. Lindsey. Monitoring and Modeling of Urban Trail Traffic: Validation of Direct Demand Models in Minneapolis, Minnesota, and Columbus, Ohio. *Transportation Research Record: Journal of the Transportation Research Board*, 2016. 2593: 47–59.

10. El Esawey, M., C. Lim, and T. Sayed. Development of a Cycling Data Model: City of Vancouver Case Study. *Canadian Journal of Civil Engineering*, Vol. 42, No. 12, 2015, pp. 1000–1010. https://doi.org/10.1139/cjce-2015-0065.
11. Lindsey, G., M. Petesch, T. Vorvick, L. Austin, and B. Holdhusen. *The Minnesota Bicycle and Counting Initiative: Institutionalizing Bicycle and Pedestrian Monitoring*. MnDOT Report No. MN/RC 2017-02. Minnesota Department of Transportation, 2017. http://www.dot.state.mn.us/research/reports/2017/201702.pdf.
12. Davis, G. A., and T. Wicklatz. *Sample-Based Estimation of Bicycle Miles of Travel (BMT)*. University of Minnesota Digital Conservancy, 2001. http://hdl.handle.net/11299/742.
13. Ryus, P., E. Ferguson, K. M. Laustsen, R. J. Schneider, F. R. Proulx, T. Hull, and L. Miranda-Moreno. *Guidebook on Pedestrian and Bicycle Volume Data Collection*. The National Academies Press, Washington, D.C., 2014.
14. Proulx, F. R., R. J. Schneider, and L. F. Miranda-Moreno. Performance Evaluation and Correction Functions for Automated Pedestrian and Bicycle Counting Technologies. *Journal of Transportation Engineering*, Vol. 142, No. 3, 2016, p. 04016002.
15. Nordback, K., D. P. Piatkowski, B. N. Janson, W. E. Marshall, K. J. Krizek, and D. S. Main. Using Inductive Loops to Count Bicycles in Mixed Traffic. *Journal of Transportation of the Institute of Transportation Engineers*, Vol. 2, No. 1, 2011, pp. 39–56.
16. Brosnan, M. Validation of Bicycle Counts from Pneumatic Tube Counters in Mixed Traffic Flows. *Transportation Research Record: Journal of the Transportation Research Board*, 2014. 2527: 80–89.
17. Hankey, S., G. Lindsey, and J. Marshall. Day-of-Year Scaling Factors and Design Considerations for Nonmotorized Traffic Monitoring Programs. *Transportation Research Record: Journal of the Transportation Research Board*, 2014. 2468: 64–73.
18. Lindsey, G., J. Wilson, E. Rubchinskaya, J. Yang, and Y. Han. Estimating Urban Trail Traffic: Methods for Existing and Proposed Trails. *Landscape and Urban Planning*, Vol. 81, No. 4, 2007, pp. 299–315. https://doi.org/10.1016/j.landurbplan.2007.01.004.
19. Wang, X., G. Lindsey, S. Hankey, and K. Hoff. Estimating Mixed-Mode Urban Trail Traffic Using Negative Binomial Regression Models. *Journal of Urban Planning and Development*, Vol. 140, No. 1, 2013, p. 04013006.
20. Turner, S., and P. Lasley. Quality Counts for Pedestrians and Bicyclists: Quality Assurance Procedures for Nonmotorized Traffic Count Data. *Transportation Research Record: Journal of the Transportation Research Board*, 2013. 2339: 57–67.
21. Miranda-Moreno, L., T. Nosal, R. Schneider, and F. Proulx. Classification of bicycle traffic patterns in five North American Cities. *Transportation Research Record: Journal of the Transportation Research Board*, 2013. 2339: 68–79.
22Nordback, K., W. Marshall, B. Janson, and E. Stolz. Estimating Annual Average Daily Bicyclists: Error and Accuracy. *Transportation Research Record: Journal of the Transportation Research Board*, 2013. 2339: 90–97.
23. Nosal, T., L. F. Miranda-Moreno, and Z. Krstulic. Incorporating Weather: Comparative Analysis of Annual Average Daily Bicyclist Traffic Estimation Methods. *Transportation Research Record: Journal of the Transportation Research Board*, 2014. 2468: 100–110.
24. El Esawey, M. Toward a Better Estimation of Annual Average Daily Bicycle Traffic: Comparison of Methods for Calculating Daily Adjustment Factors. *Transportation Research Record: Journal of the Transportation Research Board*, 2016. 2593: 28–36.
25. Jessberger, S., R. Krile, J. Schroeder, F. Todt, and J. Feng. Improved Annual Average Daily Traffic Estimation Processes. *Transportation Research Record: Journal of the Transportation Research Board*, 2016. 2593: 103–109.
26. Gobster, P., S. Sachdeva, and G. Lindsey. Up on the 606: Understanding Use of a New Elevated Pedestrian and Bicycle Trail in Chicago, Illinois. *Transportation Research Record. Journal of the Transportation Research Board*, 2017. 2644: 83–91.
27. Wilson, J., and G. Lindsey. *Design and Implementation of a Method for Selecting Sampling Locations for Trail Traffic Monitoring*. Prepared for Rails to Trails Conservancy, Inc., Yellowsprings, Ohio, 2016.

The Standing Committee on Highway Traffic Monitoring (ABJ35) peer-reviewed this paper (18-06289).

Article

TRR
JOURNAL OF THE TRANSPORTATION RESEARCH BOARD

Transportation Research Record
2018, Vol. 2672(43) 110–121

DOI: 10.1177/0361198118794048
journals.sagepub.com/home/trr
SAGE

Methods for Estimating Axle Factors and Axle Classes from Vehicle Length Data

Raul E. Avelar[1], Scott Petersen[2], Tomas Lindheimer[1], Sruthi Ashraf[1], and Erik Minge[2]

Abstract

This study developed methods to estimate axle factors and vehicle class from length-based data streams. A set of eight methods was proposed and evaluated in different testing schemes intended to observe performance on homogeneous and heterogeneous data. The initial analysis used length-based data from 61 sites in Wisconsin. The research team compared performance of the methods estimating axle factors and vehicle class proportions. Performance was comparable and consistent between homogeneous and heterogeneous subsets of data. The research team selected two methods for a final round of analysis based on their accuracy and robustness to heterogeneity. For the final round of analysis, the research team assembled a multistate dataset using data from Wisconsin and from 14 other states represented in a dataset from the Long Term Pavement Performance program. The final round of analysis compared performance under different seasons, facility type, and road character (urban vs. rural). Performance of the two identified methods was deemed appropriate and they are recommended for implementation.

There are many important uses of vehicle classification data. For example, to calculate annual average daily traffic (AADT) for a stream of axle counts, one needs to have an estimate of the axle factor. An axle factor (AF) is the ratio between the total number of vehicles to the total number of axles at that location. The AADT is simply the product between the count of axle hits and the AF. Another use of vehicle classification data is to estimate pavement loads. Accurate and consistent load estimates derived from traffic data are critical inputs for performing pavement design and assessment of its performance. In the 1980s, FHWA developed a standardized 13-category vehicle classification scheme. The following list shows the 13 FHWA classes and a brief description of each (*1*):

- Class 1 Motorcycles; 2 axles, 2 or 3 tires.
- Class 2 Passenger cars; 2 axles, can have 1-or 2-axle trailers.
- Class 3 Pickups, panels, vans; 2 axles, 4-tire single units, can have 1 or 2 axle trailers.
- Class 4 Buses 2 or 3 axles, full length.
- Class 5 Single unit 2-axle trucks; 2 axles, 6 tires (dual rear tires), single unit.
- Class 6 Single unit 3-axle trucks; 3 axles, single unit.
- Class 7 Single unit 4 or more-axle trucks; 4 or more axles, single unit.
- Class 8 Single trailer 3- or 4- axle trucks; 3 or 4 axles, single trailer.
- Class 9 Single trailer 5-axle trucks; 5 axles, single trailer.
- Class 10 Single trailer 6 or more-axle trucks; 6 or more axles, single trailer.
- Class 11 Multi-trailer 5 or less-axle trucks; 5 or less axles, multiple trailers.
- Class 12 Multi-trailer 6-axle trucks; 6 axles, multiple trailers.
- Class 13 Multi-trailer 7 or more-axle trucks; 7 or more axles, multiple trailers.

The use of FHWA vehicle classes is widespread in the U.S. and thus, the estimation of the number of vehicles in each class is a key task that transportation agencies must undertake routinely. The methods to obtain such estimates naturally depend on the types of sensors used.

[1]Texas A&M Transportation Institute, College Station, TX
[2]SRF Consulting Group, Minneapolis, MN

Corresponding Author:
Address correspondence to Raul E. Avelar: r-avelar@tamu.edu

Such AF and vehicle classification estimates have been derived from in-pavement axle sensors that are more costly to deploy and maintain than other sensors, particularly non-intrusive length sensors. Such non-intrusive sensors are now widespread and generate accurate traffic data. It is important to be able to understand whether these estimation tasks can be performed from the data from length-based sensors.

Research Objective

The objective of this research is to investigate and document the performance of eight methods for estimating AFs and the FHWA axle-based vehicle classes using binned vehicle length data.

Literature Review

Currently, there are two categories of widely used vehicle classification schemes: axle based or length based. The most prominent axle-based classification technologies are in-roadway sensors, including piezoelectric sensors, weigh-in-motion scales, infrared sensors, and specialized inductive loop systems (*2*).

The Long Term Pavement Performance (LTPP) program has developed and deployed a set of rules for converting axle spacing and weight data into vehicle classification data, which are being used by Transportation Pooled Fund Study (TPF) weigh-in-motion (WIM) sites across the country (*3*). The LTPP classification scheme uses a combination of four variables to classify each vehicle (*3*):

- Number of axles on the vehicle,
- Spacing between those axles,
- Weight of the first axle of the vehicle, and
- Gross vehicle weight of the vehicle.

The LTPP produced different results when compared to different state-developed classification schemes, resulting in different classification results.

Estimation from Pavement Sensors

Various algorithms and rules exist that convert axle spacing data from inductive loops and data from weigh sensors into vehicle classification estimates for a given site. Such rules and algorithms vary from state to state and from vendor to vendor (*3*). As the number of axles and their spacing characteristics varies by vehicle within each category, the task of classifying vehicle data into the 13 FHWA classes is not a simple task.

When using pavement-based sensors, transportation departments optimize the classification rules to their state's vehicle mix by shifting their break points. Geographical transferability of such classification systems is not always achievable and the development of additional rules is necessary to help detect and monitor specific vehicles that are important to each State (*3*).

WIM sensors use both axle spacing and axle weight data to classify vehicles whereas conventional vehicle classifiers lack weight data, which may result in the same vehicle being classified differently. Commonly used WIM sensors are piezoelectric, load cell, and bending plate systems (*2*). WIM sensors measure dynamic axle weights which are different from static axle weights. Loop detectors work well in free-flow conditions based on negligible vehicle acceleration assumption (*4*) and have the ability to detect lifted axles of heavy goods vehicles under normal traffic conditions, enhancing their proper classification (*5*).

Axle Factors

To convert an axle count to AADT volume, an AF is used. AFs are calculated by analyzing axle-based classification data and determining the ratio of total vehicles to the axle count. Data from such axle-based vehicle classification sites are significantly costly. Length-based vehicle classification (LBVC) sensors have become prevalent, as their installation and operation is significantly cheaper (*6*).

LBVC sensors can be placed on, above, or beneath the roadway. LBVC sensors are often more flexible to install, replace, and maintain compared with axle-based technologies (*7*); however lower accuracy is a concern for some sensors (*8*).

Length-based detectors include loop detectors and several types of non-loop sensors such as sidefire and in-road sensors. Currently, loop detectors are the most frequently deployed detection systems (*9*). Video detection systems and continuous-wave Doppler sensors are examples of other types of LBVC sensors. A major disadvantage of LBVC technology is that it does not necessarily match well with the 13-vehicle category classification system. Typically, the lengths of vehicles are aggregated into length bins by the sensor. However, the lengths of the vehicles in the 13 categories tend to overlap. A three- or four-bin classification system is used by most agencies in order to differentiate passenger cars from trucks (*2*). Some states may add categories to FHWA's vehicle classification and often set their own bin-length definitions to better differentiate the types of vehicles from their jurisdictions.

Vehicle Classification. Distribution of vehicle types (*10*) and axle spacing characteristics (*2*) can significantly vary spatially and temporally. There can be errors in axle-based classification caused by incorrect grouping of vehicles (*11*). However, there is no standard scheme for LBVC, and data are often aggregated into three to five length-based bins.

In their study using LIDAR, Lee and Coifman developed a cost-effective method to check accuracy of vehicle classification stations. The researchers classified vehicles to six categories using vehicle length, height and middle drop details to monitor permanent and temporary vehicle classification stations, including axle- and length-based classification stations (*12*).

Liu et al. developed a multi-layer perceptron neural network approach and equipped existing inductive loops with advanced high-scanning rate detector cards to estimate vehicle emissions. Vehicle emissions were estimated from the vehicle mix and vehicle activity data obtained from the inductive loops (*13*). The classification system had a correct classification rate (CCR) of 90 percent. In another study by Oh and Ritchie, a probabilistic neural network multivariate Bayesian classification scheme and a heuristic classification algorithm which made use of inductive signatures produced overall CCR of 70.8 percent (*14*).

Coifman used estimated speeds from single-loop measurements to estimate vehicle lengths. He found that the major difficulties with LVBC are the common instances of inoperable or malfunctioning detectors rather than the measurement scheme (*15*). Coifman emphasizes the need for continuous monitoring of classification sensors. Zhang et al. used a three-layered artificial neural network to classify vehicles into bins (*16*). The authors argue that deterministic mathematical equations cannot capture stochastic features of traffic flow completely.

Coifman and Kim classified vehicles into three length bins with divisions of effective vehicle lengths at 28 feet for passenger cars and 46 feet for large trucks, using freeway single-loop detector data to differentiate between large trucks and passenger cars even in congestion and with high percentage of long vehicles (*17*). Over 97% of the vehicle lengths were correctly classified into the three length bins, with over-classification (estimated length bin higher than the measured length) being prevalent. When compared with concurrent dual-loop detector and video data, length-based single-loop detector length-estimation results were very close to those from the dual-loop detector measured length classification results, after excluding pulse-breakup.

Minge et al. recommended a four-bin-length scheme with a five-bin scheme for areas with a significant number of long combination vehicles and with anticipation of overlap at the edge of class bins (*9*). Locational and temporal influences were found to vary the optimal value of bin boundaries (*18*).

Five-length-based bins (which can be collapsed to four-length bins) and six axle-based classes, with the aid of per-vehicle records (PVR) data to estimate number of axles per bin, were used to contain all 13 classes of vehicles outlined by the FHWA axle-based classification (*8*). Little variation was seen across states or years in correlation of axle classification to length classification. A vehicle classification strategy was developed strictly based on measured effective vehicle length and measured speed to improve length-based classification during congestion to identify impacts of the unobserved acceleration on the measured length class (*4*). Liu and Sun developed an algorithm based on Newell's simplified car-following model that could classify vehicles passing by a loop detector into two categories, long vehicles and regular cars (*19*). The algorithm made use of event-based loop detector data that contained every vehicle detector actuation and de-actuation event.

Axle Factor Estimation. Many states use AF to adjust single tube traffic counts. The Ohio Department of Transportation advises that AFs should be used cautiously when irregular traffic patterns are observed (near major recreational areas, etc.). AFs for each functional class of road (both urban and rural) based on 24- and 48-hour portable vehicle classification counts were found to vary temporally (*20*). Load spectra obtained from LBVC counts provide a good representation of actual loads incurred on the road, even though Minge et al. found that these spectra are less accurate than axle-based spectra (*9*).

As length-based AFs have more variability because of a changing number of axles per bin over time, Forde and Wilkinson recommended supplemental software (*8*) that uses PVR to measure changes in the average number of axles per bin to aid length classification to be a valid source of AF input.

From the review of previous studies, the research team noted that LBVC systems are viable alternatives to in-pavement systems and potentially offer flexibility and cost savings (*7, 8, 21*). It was found that several studies have been conducted to accurately estimate AFs and to classify the vehicles into the 13 FHWA vehicle class categories. Studies suggest that binning of vehicles based on length for vehicle classification purposes may provide results similar to axle-based vehicle classification but the results vary with method adopted for classification and number of length bins and corresponding thresholds.

Data Characteristics

This study used two different datasets to perform various assessments. All data included both vehicle length and axle-based vehicle classification data.

Wisconsin Dataset

The dataset from Wisconsin included data from 61 sites and consisted of over 490 million individual PVR with information on both vehicle length and vehicle class. The data was binned prior to the analyses. Vehicles with length between 15 and 24 feet represented about 90% of the records. Most of the rest of the data came from vehicles with lengths ranging from 5 to 14 feet. Around 1.6 million records (or 0.3% of the dataset) were categorized as "unknown." These were most likely vehicles with axle spacing longer than 40 feet, probably Class 9 vehicles with odd axle spacing. This category of unknown class vehicles was treated as a 14th class for the purposes of this research. The counts for vehicles of classes 5, 6, 7, and 8 were, respectively, 10.2, 2.3, 0.9, and 5.8 million records, with a majority of vehicle lengths ranging from 15 ft to 74 ft.

LTPP Dataset

Transferability and performance of the selected methods developed from the initial analysis were evaluated using the complete data sets from the 17 additional sites provided by the LTPP. The sites were located in Arizona, Arkansas, California, Colorado, Illinois, Indiana, Kansas, Louisiana, Maryland, Minnesota, New Mexico, Pennsylvania, Tennessee, Virginia, and Wisconsin. These data had the same characteristics of the Wisconsin dataset, with all 14 classes known (13 FHWA classes plus a class of vehicles of "unknown class").

Study Design

Researchers developed a two-phase testing plan for an initial round of analyses based on a subset of the data. The two-phase testing plan used the 61 sites provided by Wisconsin. Before the initial set of analyses, the research team identified 17 sites from Wisconsin with the proportion of vehicle classes within the typical range for the data set (that is, with all their class proportions within the 10th and 90th percentiles of each proportion). One subset of seven sites, selected randomly from these 17, was used to calibrate the methods of analysis. The remaining 10 sites were reserved for evaluation of the methods. The second phase tested the performance of the methods when applied to data from the remaining 44 sites, a more heterogeneous subset of sites that includes some deviations in the traffic composition, compared with the calibration subset.

The calibration subset consisted of 106,803,612 records with "average" proportions of vehicles, per the described preliminary assessment of the whole dataset. The dataset for the first phase testing on 10 typical sites contained 94,472,209 individual records. In total, 289,880,973 records from 44 sites were available for the second phase tests.

For the second and final round of analysis, data from the LTPP data set was combined with the Wisconsin data in order to assess the performance of the two most promising methods of estimation identified in the first round of analyses.

Methods of Estimation

For this research, eight methods were proposed and their performance examined. All these methods rely on a calibration procedure using a subset of data where both length and axle class (AC) are known (termed the "calibration data" throughout the paper).

Method 1 used the calibration dataset with PVR AC data to determine typical numbers of axles per bin and generated an AF. The steps for this method are shown next:

1. Retrieve calibration data.
2. Determine ideal bin thresholds based on natural breaks in the calibration dataset.
3. Calculate of the average number of axles per length bin from calibration data.
4. Retrieve the vehicle count per optimal length bin from a dataset for analysis (binned length data).
5. Apply the average number of axles to vehicle counts per length bin in the dataset for analysis and calculate the AF.

Method 2 has the same steps for method 1 except that the calibration steps (steps 1 and 2) are performed using representative sites only. In contrast, compared with method 1, the calibration step is performed for each site, assuming that a representative sample of PVR data can be obtained from each site.

Method 3 is similar to method 1, except a set of predetermined length bins is used instead of determining an optimal set of bins.

Similar to method 3, method 4 uses predefined length class (LC) bins instead of performing step 2 in the list above. The difference between these methods is analogous to the difference between methods 1 and 2: for method 3, it is assumed that the average number of axles per vehicle in each bin is performed with PVR data from every site; this determination is only made once, using a set of representative sites, for method 4.

In contrast with the previous methods, Method 5 used PVR AC data (AC and length) from a calibration dataset to establish probability density functions for vehicle lengths per AC. This set of functions, along with a set of optimal bin thresholds (manually determined as for method 1) and binned dataset for analysis, are the inputs

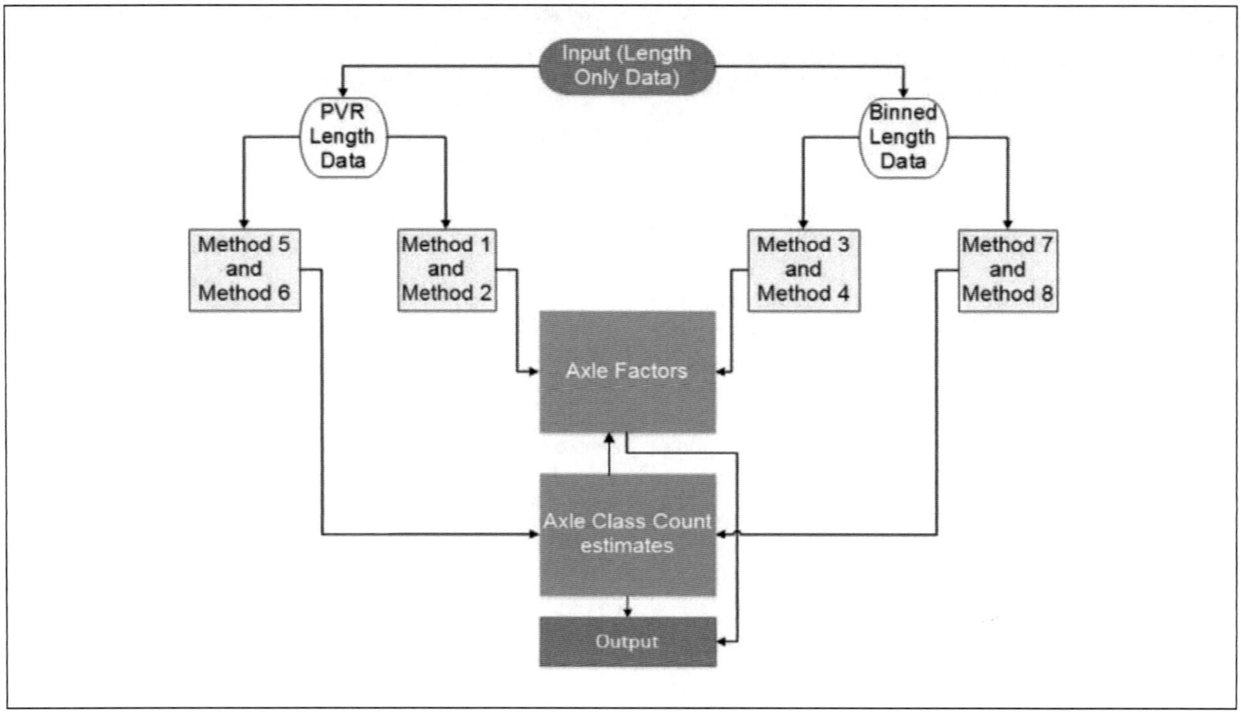

Figure 1. Schematic of methods, inputs and outputs.

of an algorithm that estimates a parametric function of the mix of vehicle classes at the site.

Method 6 utilized the calibration data to determine a fractional assignment scheme per bin. This scheme is then used to classify the binned vehicle data into ACs.

The steps for method 7 are the same as for method 5, except that the binned LC data and set of bin thresholds are predetermined (as in method 3).

Method 8 was similar to method 6 but performed fractional assignment of binned LC data to AC. Figure 1 shows a schematic of the methods, their inputs and their outputs.

The outputs of methods 1, 2, 3, and 4 were AFs, and the outputs for methods 5, 6, 7, and 8 were AC count estimates. Methods 5 and 7 are parametric methods in essence (i.e., an algorithm that uses probability distributions of vehicle lengths to categorize them into ACs), whereas methods 6 and 8 were fractional assignment methods (simpler algorithm for factoring vehicle lengths into ACs).

In order to obtain AFs from any of these methods, the number of axles per vehicle class from calibration data must be known. Based on the different sets of length thresholds from each method, specific calibration values were obtained from the calibration dataset to use when the methods were applied to the test datasets. As the calibration subset was the same for all methods, differences in calibration outcomes pointed to the sensitivity of each method to the choice of the thresholds of aggregation for the bins. Two sets of four-length bins were considered to evaluate the methods.

Metrics for Performance for Comparative Analysis

Based on the Wisconsin dataset, phase 1 of the initial round of analyses utilized data from 10 "typical" sites, whereas phase 2 used data from 44 sites that exhibited heterogeneous distributions of vehicle classes. The final round of analyses used data from the 17 LTPP sites in combination with the 54 sites from the initial round of analyses. The research team contrasted the performance of these methods based on the metrics summarized in Table 1.

For the second round, only the methods that produced the best results in estimating AFs and AC count estimates were used. The purpose of the second round of analyses was to perform a robust evaluation of accuracy considering several factors of interest including: season, facility type (freeway or highway), number of lanes, and whether the site was in an urban or rural location.

Analysis and Results

This section summarizes the analyses the research team performed, using the methods and metrics described in the previous section, and the corresponding results.

Table 1. Metrics of Performance for Comparative Analysis

Performance measure	Analysis phase	Metric	Description	Units of analysis
Axle factor	Initial analysis (phase 1 and phase 2) and final analysis	Relative error	The average of the deviation of the estimated AF per unit of analysis (in percent) relative to the true AF	Per site, per test
		Absolute error	The average of the absolute deviation of the estimated AF per unit of analysis (in AF units)	Per site, per test
Proportion of vehicles per class	Initial analysis (phase 1 and phase 2)	Relative error	The average of the deviation of the estimated Vehicle Class proportion per unit of analysis (in percent) relative to the true proportion	Per class, per site, per test
		Absolute error	The average of the absolute deviation of the estimated Vehicle Class proportion per unit of analysis (in AF units)	Per class, per site, per test
	Final analysis	Proportion of vehicles misclassified	The ratio of the cumulative deviation of the estimated Vehicle Class proportion per unit of analysis (in number of vehicles) to the total number of vehicles	Per site, per test

From these evaluations, methods 1 and 2 produced exactly the same results as did methods 3 and 4. The research team estimated the required number of axles per bin only once in the calibration phase to avoid a manual examination per sites other than the seven sites in the calibration data. This modification is expected to yield more conservative results because the performance is expected to improve if a set of optimal bins were determined per site. However, the methods would be more attractive for implementation if the performance were acceptable when only using a subset of data for calibration.

Initial Round, Phase 1 Analyses

All methods performed with a relative error less than 1% in AF estimation. Error from method 3 was over four times as large as the error from method 1, which suggests sensitivity to the choice of bins. Figure 2 shows the AF estimation performance of all methods during phase 1 of the initial round of analysis.

Relative error and absolute error in estimation of number of vehicles by vehicle class was compared for methods 5, 6, 7, and 8. On average, all methods performed within 0.3% absolute error. The best performing methods in recognizing vehicle classes were methods 5 and 6. These methods were obtained based on bins with thresholds at 7 feet, 27 feet and 53 feet. The mean absolute error in vehicle class proportion doubled for other two methods that rely on bins with thresholds at 7 feet, 19 feet, and 33 feet (i.e., methods 7 and 8). The highest classification errors occurred for classes 11 and 12. Despite a very comparable performance of all methods over the 14 classes, the two parametric methods (methods 5 and 7) performed better than the two fractional assignment methods in estimating the proportion of two of the most prominent classes in the vehicle mix, classes 2 and 9.

Initial Round, Phase 2 Analyses

The research team compared the performance of methods 1 through 8 using the data from the remaining 44 sites. All methods performed with relative errors less than 1% except for methods 3, 4, and 8, which had relative errors of about 1.7%. Choice of bin thresholds affected the magnitude of the estimation error for non-parametric methods (1 vs. 3 and 6 vs. 8) at a higher rate for their parametric counterparts. Figure 3 shows the AF estimation performance for phase 2 tests.

All four methods performed worse in phase 2 than in phase 1. This was expected because the phase 2 dataset has more heterogeneous class proportions than phase 1 sites. Again, the parametric methods 5 and 7 performed better than the fractional assignment methods identifying classes 11 and 12, which had highest relative errors. The two fractional assignment methods were equally robust to the two parametric methods regarding the choice of bins. Parametric methods had smaller absolute errors in recognizing the proportions of classes 2 and 9, but non-parametric methods performed better recognizing classes 3 and 4.

The research team performed an additional set of assessments to compare the changes in errors between the sets of analysis when the only difference is the heterogeneity of the data so that it may help anticipate the change in performance when using a larger dataset. When applied to heterogeneous data, the performance in AF estimation was generally worse. All methods experienced an increase in errors with respect to their errors

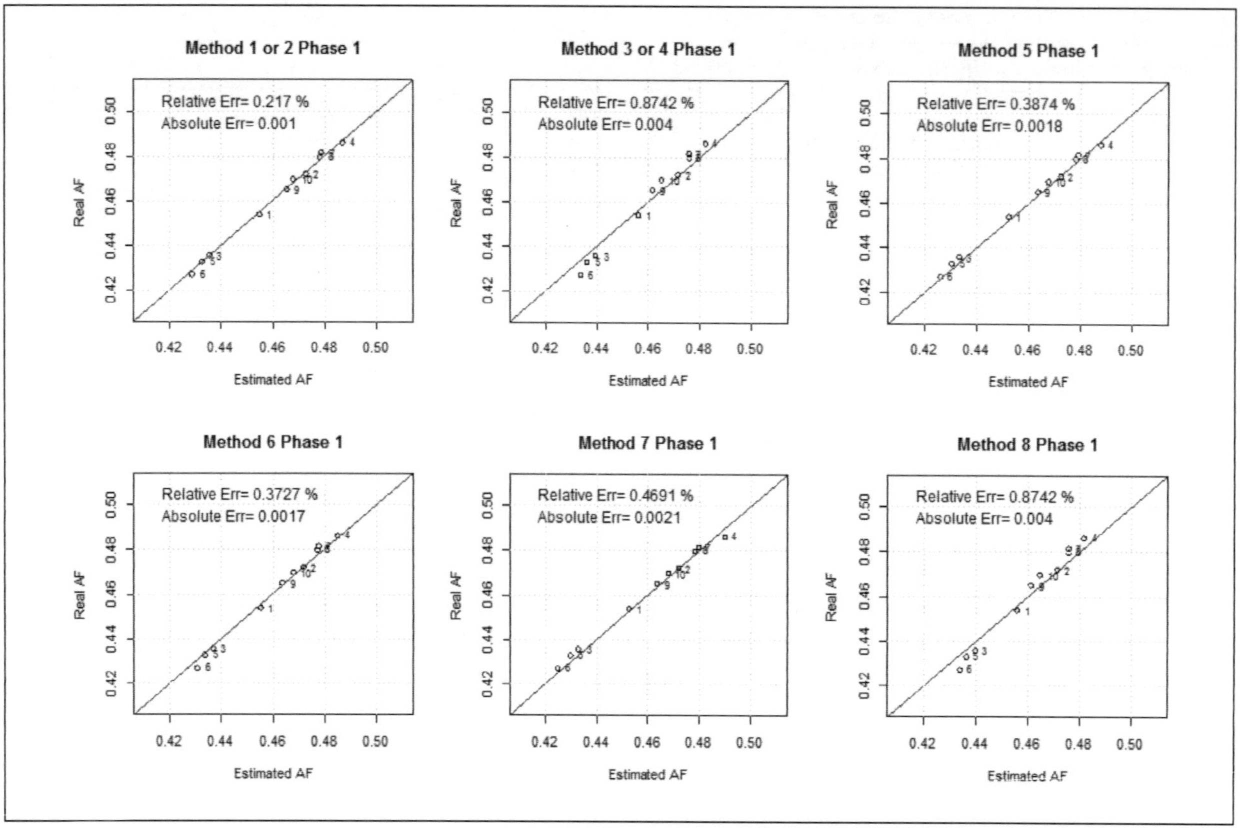

Figure 2. Axle factor estimation performance of methods for phase 1 tests.

Table 2. Ratios in Absolute Error

Method	Phase 2/ Phase 1 absolute error ratio (for AFs)	Phase 2/ Phase 1 absolute error ratio (for class proportions)
1 and 2	1.90	NA
3 and 4	1.95	NA
5	1.66	4.32
6	2.34	8.06
7	1.49	4.77
8	1.95	6.78

Note: NA = not applicable.

when using homogeneous data, but remained within acceptable boundaries. The assessments are shown in Table 2. Methods 5 and 7 were most resilient to the increase in data heterogeneity when estimating both AFs and proportions of vehicle. Method 1 performed better than method 3 when the objective was only AF estimation.

Based on phase 1 and phase 2 analyses, method 1 and method 5 were selected for the last round of analyses. Method 1 is recommended for its best estimation of AFs and overall simplicity, if there is no need to estimate ACs. Method 5 is recommended when ACs are required. Method 5 produced: (1) smaller relative errors in class proportions overall; (2) smaller absolute errors for two of the key classes of vehicles: classes 2 and 9; and (3) robustness against heterogeneity and choice of bins in the data compared with other competing methods that also estimate vehicle classes.

Final Round, Combined Datasets Analyses

Analyses were conducted based on the combined data from Wisconsin and LTPP. More nuanced differences in season, type of facility, road character, and number of lanes were also made in this round of analyses.

Axle Factor Estimation Performance. As shown in Figure 4, Method 1 produced better results in AF prediction with a relative error of 0.67% on heterogeneous data and method 5 having slightly worse performance

Regardless of road character, type of facility, and cross-section the following observations emerge from these analyses:

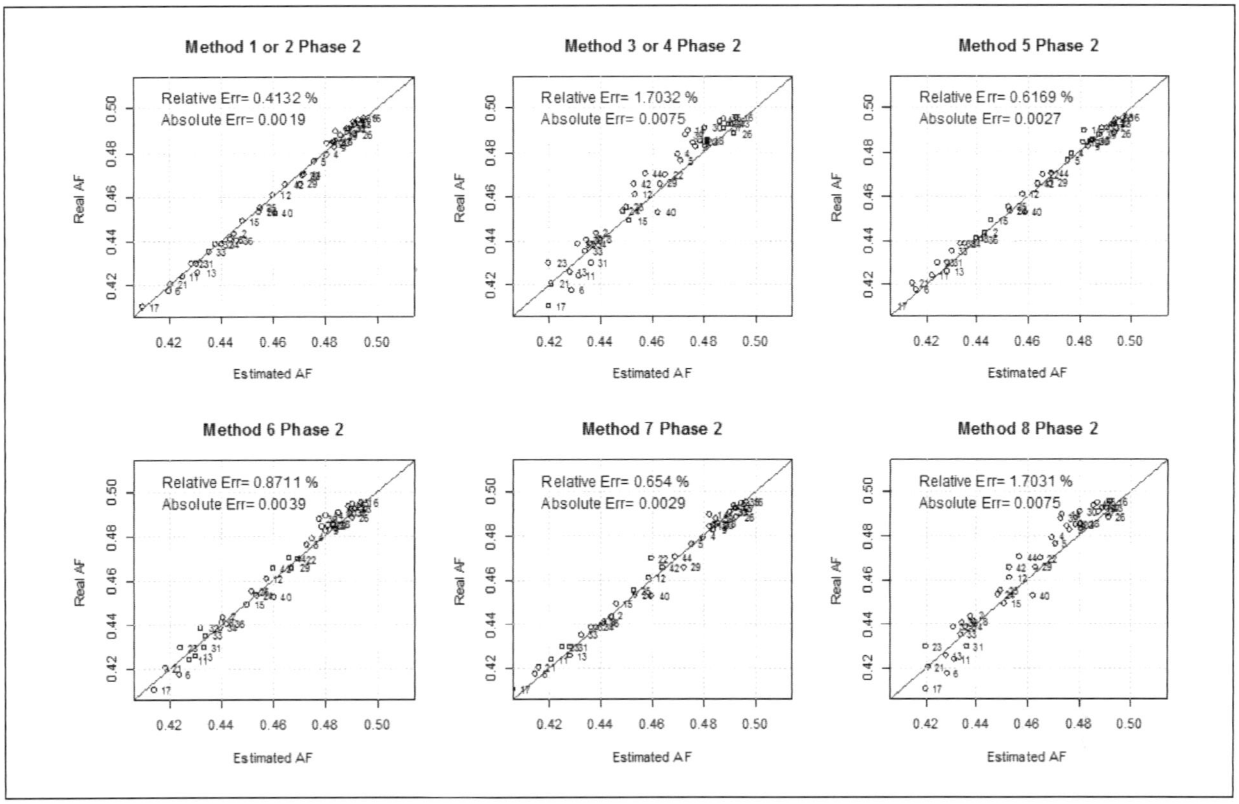

Figure 3. Axle factor estimation performance of methods for phase 2 tests.

- Method 1 tends to slightly overestimate the true AF at low values of AF.
- Method 5 tends to slightly underestimate the true AF at low values of AF.

Even with the trend of slight bias, the predicted AF remains within 2% of the true AF value. AF distribution along urban roads was skewed toward small values and had a slightly higher median than AF values of rural roads. Highways tend to have a compact distribution of AFs with larger values, compared with freeways. Two-lane roads showed higher AFs compared with four-lane roads, where the values tend to be lower and span a wider range at these facilities.

Vehicle Class Estimation Performance. The research team performed tests to assess the performance of method 5 regarding its ability to recognize vehicle classes correctly by determining the proportion of vehicles misclassified. In this battery of tests, the research team observed variability in the proportion of misclassifications as a performance metric.

To provide an overall depiction of performance, the research team plotted real versus estimated proportions of vehicles for each class, as shown in Figure 5. The 1:1 line is shown in blue for comparison, which indicates an idealized perfect correlation between prediction and real proportion. Accuracy of classification was highest for classes 1, 2, 8, 9 and for the vehicles of unknown class.

In general, the proportion of misclassifications tended to be independent of the AF value with median roughly static at about 10% for this round of tests. The only interesting trend was perhaps a smaller proportion of misclassified vehicles at urban sites compared with rural sites, and on freeways compared to highways, but to a lesser extent than the contrast between rural and urban facilities.

When only Wisconsin data are considered, the performance of recognizing classes 1, 8, and 9 were the highest. The classes that seem to have less precision were classes 4, 13, and vehicles of unknown class. Analysis with only LTPP data estimated the proportion of classes 2 and 9 vehicles the best, but performance was marginal for almost every other class. This reduced performance is the result of differences between the Wisconsin and LTPP datasets, and it is expected that performance would improve significantly if the algorithm were recalibrated using LTPP data as well as Wisconsin data.

Seasonal Variability. The research team subdivided the data into fall (months of October and November), winter

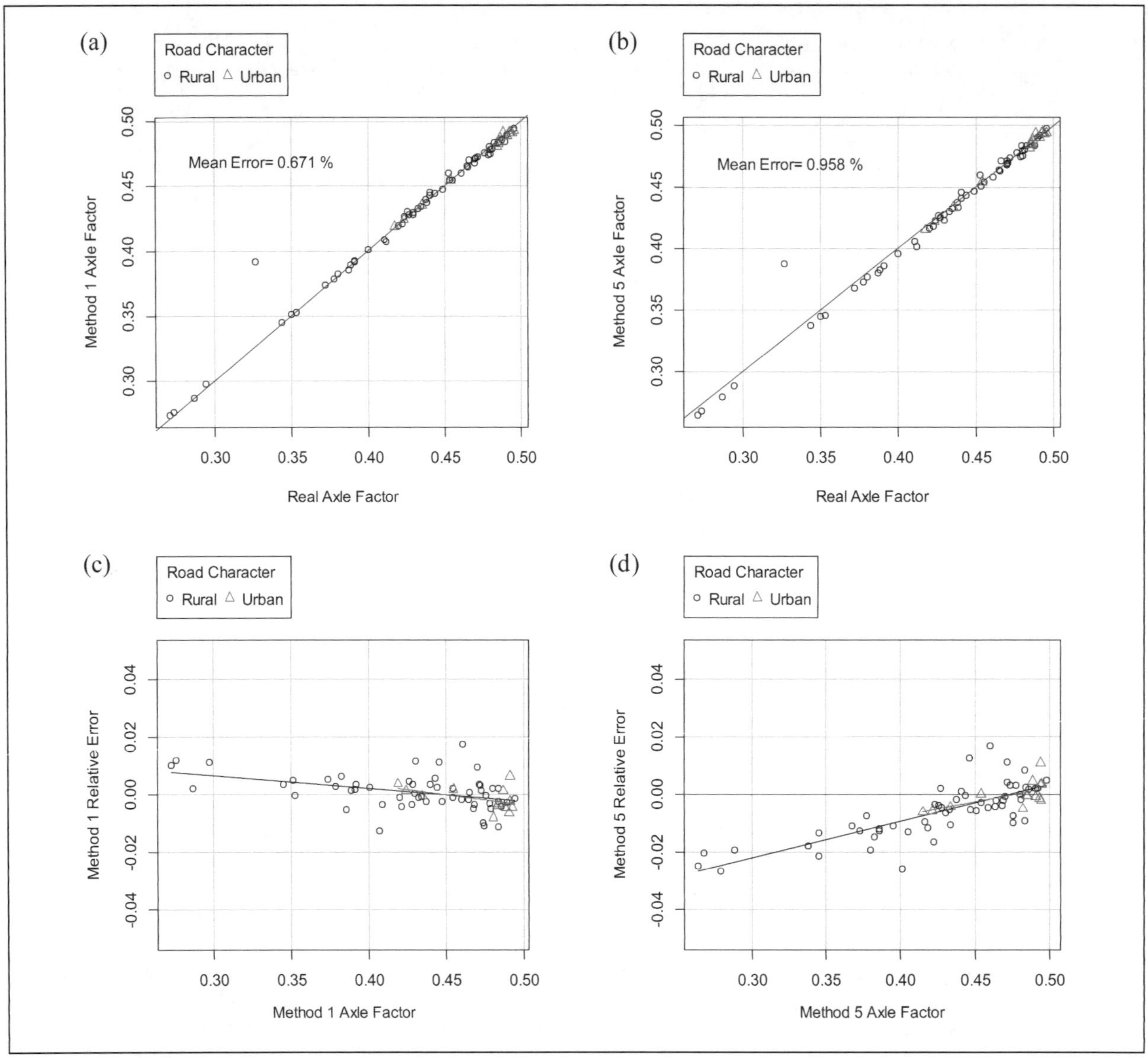

Figure 4. Comparison of axle factor estimation by method 1 and method 5; (*a*) method 1; (*b*) method 5; (*c*) method 1 relative errors; (*d*) method 5 relative errors.

(January and February), and summer (July and August) to examine seasonal differences in performance. No trends were apparent when subdividing the data this way, an indication of robustness against seasonal fluctuations in traffic.

Conclusions and Recommendations

The objective of this paper is to develop and test methods to estimate AFs and vehicle classes using length data. As LBVC technologies are economical and flexible in installation, maintenance, and replacement, the methods discussed used length-based data only to develop AFs. The eight methods were calibrated and evaluated using data from the state of Wisconsin. The performance of the best performing methods was then evaluated and verified using data from Wisconsin and LTPP.

Using the results of the tests over the Wisconsin dataset, the research team recommends methods 1 and 5, given their superior performance. Methods 1 or 2 performed the best in estimating AFs and are recommended if there is no need to estimate ACs. If ACs are required, then method 5 is recommended. In addition, method 5 showed robustness against the choice of bins, as well as the heterogeneity in the data.

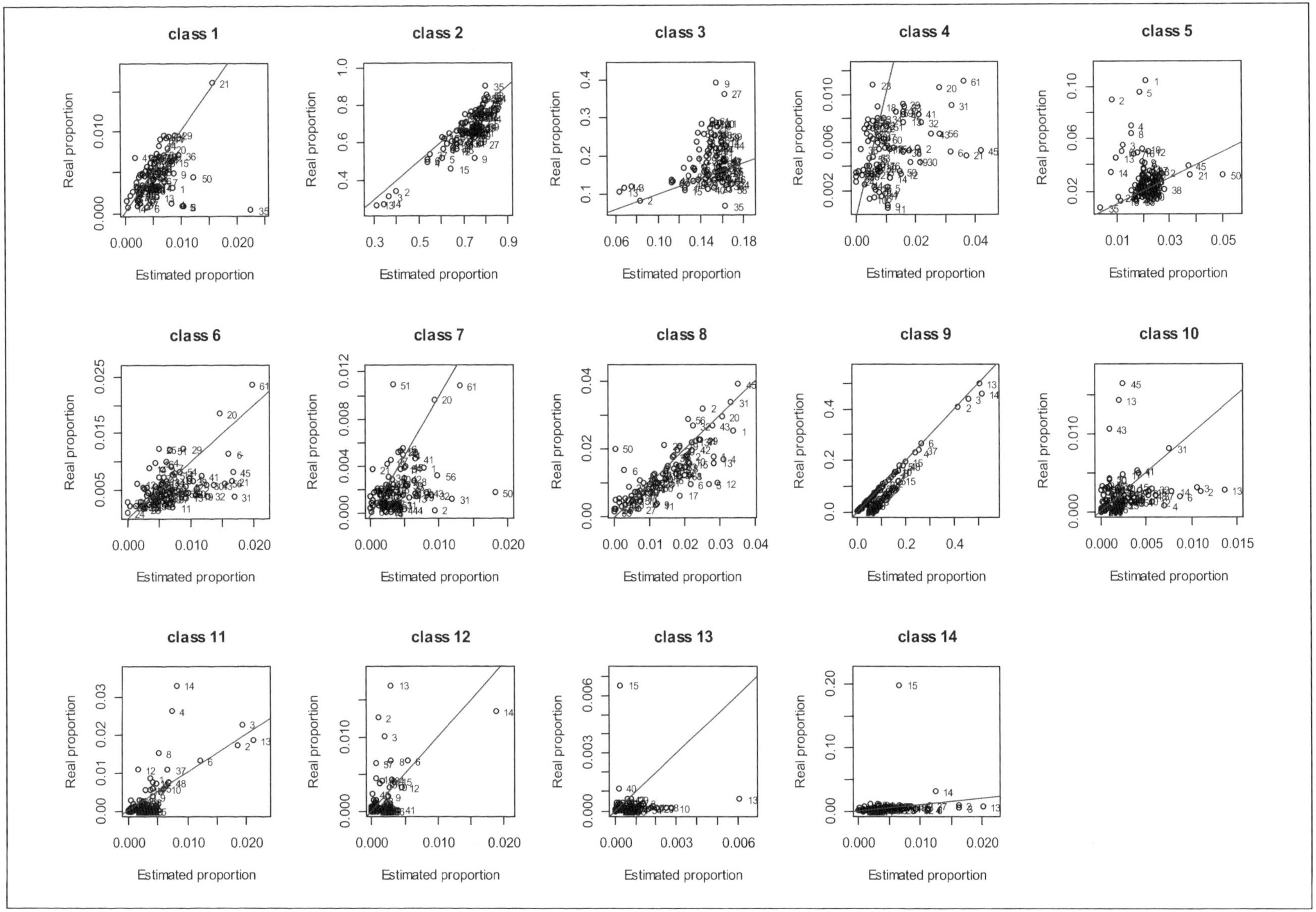

Figure 5. Real and estimated proportion of vehicle classes for Wisconsin and LTPP dataset.

The research team assembled a multistate dataset using Wisconsin and LTPP data for a final round of performance tests on methods 1 and 5. Based on the results of the final round of evaluations, the research team found performance of the two recommended methods within the expected tolerance limits, as suggested by the first round of analysis. The evaluations based on AF prediction confirmed the performance of these methods to be within 2% of the predicted AF value. The research team identified a trend to over predict AF at low AF values for method 1 and the opposite trend for method 5. However, none of these trends exceeded acceptable amounts of error. The trends were invariant when looking at subsets of data by road character, facility type, and number of lanes. This invariability suggests that a simple post-hoc adjustment of the predicted values produced by either method may help control for the bias and reduce the overall error.

Based on the systematic trend of the bias, the research team recommends developing a post-hoc adjustment in the prediction of AFs and incorporating such adjustment in the implementation of this method. However, given the small magnitude of the bias observed, acceptable performance is expected even if no adjustment is developed at all.

Regarding the prediction of vehicle classes by method 5, the evaluations showed a median misclassification proportion of about 10%, regardless of differences between data sources, road character, facility types, and season. The proportion of misclassification was minimal at urban sites with lower AFs and at freeways with higher AFs. Similar to the error trends in AF estimation, these trends have the potential to inform a post-hoc adjustment of the results of estimates from method 5.

Acknowledgments

The authors thank Russell Lewis from Wisconsin Department of Transportation and the rest of the Transportation Pooled Fund Program Advisory Panel for funding this research and for their insights, assistance, and for providing the data for this research. More details about this research are available in the project report Viability of Vehicle Length in Estimating Vehicle Classification and Axle Factors (TPF-5(340)). The authors also thank Michael Janson, George Gillette, and other researchers at TTI and SRF for their assistance throughout this research project.

Author Contributions

The authors confirm contribution to the paper as follows: study conception and design: Raul Avelar; data collection: Scott Petersen, Sruthi Ashraf, Tomas Lindheimer, and Erik Minge; analysis and interpretation of results: Raul Avelar and Scott Petersen; draft manuscript preparation: Raul Avelar, Tomas Lindheimer, and Sruthi Ashraf. All authors reviewed the results and approved the final version of the manuscript.

References

1. *Traffic Recorder Instruction Manual.* Texas Department of Transportation, Austin, TX, 2012. http://onlinemanuals.txdot.gov. Accessed July 3, 2017.
2. *Traffic Monitoring Guide.* Federal Highway Administration, Washington, D.C., 2016. www.fhwa.dot.gov/policyinformation/tmguide/tmg_fhwa_pl_17_003.pdf. Accessed April 24, 2017.
3. Hallenbeck, M. E., O. I. Selezneva, and R. Quinley. *Verification, Refinement, and Applicability of Long-Term Pavement Performance Vehicle Classification Rules.* Publication FHWA-HRT-13-091. Federal Highway Administration, U.S. Department of Transportation, Washington, D.C., 2014.
4. Wu, L., and B. Coifman. Improved Vehicle Classification from Dual-Loop Detectors in Congested Traffic. *Transportation Research Part C- Emerging Technologies*, Vol. 46, 2014, pp. 222–234. http://doi.org/10.1016/j.trc.2014.04.015.
5. Gajda, J., P. Piwowar, R. Sroka, M. Stencel, and T. Zeglen. Application of Inductive Loops as Wheel Detectors. *Transportation Research Part C- Emerging Technologies*, Vol. 21, 2012, pp. 57–66. http://doi.org/10.1016/j.trc.2011.08.010.
6. *AASHTO Guidelines for Traffic Data Programs.* American Association of State Highway and Transportation Officials, Washington, D.C., 2009.
7. Zhang, L., H. Carl, and L. S. Tighe. *Evaluating Weigh-In-Motion Sensing Technology for Traffic Data Collection.* Annual Conference of the Transportation Association of Canada, Saskatchewan, Canada, 2007, pp. 1–18.
8. Forde, S., and L. Wilkinson. *Understanding the Significance of Axle Versus Length Classification on Axle Factors and the Effect on AADT to Ensure Reliable Traffic Data.* North American Travel Monitoring Exposition and Conference, Chicago, IL, 2014. http://onlinepubs.trb.org/onlinepubs/conferences/2014/NATMEC/Forde.pdf. Accessed September 13, 2016.
9. Minge, E., S. Peterson, H. Weinblatt, B. Coifman, and E. Hoekman. *Loop- and Length-Based Vehicle Classification.* Publication MN/RC 2012-33. SRF Consulting Group, Minnesota Department of Transportation, Saint Paul, MN, 2012.
10. Malcom, C., T. Younglove, M. Barth, and N. Davis. Mobile-Source Emissions: Analysis of Spatial Variability in Vehicle Activity Ppatterns and Vehicle Fleet Distributions. *Transportation Research Record: Journal of Transportation Research Board*, 2003.1842: 91–98.
11. Minge, E., J. Kotzenmacher, and S. Peterson. *Evaluation of Non- Intrusive Technologies for Traffic Detection.* Publication MN/RC 2010-36. SRF Consulting Group, Minnesota Department of Transportation, Saint Paul, MN, 2010.
12. Lee, H., and B. Coifman. Using Lidar to Validate the Performance of Vehicle Classification Stations. *Journal of Intelligent Transportation Systems*, Vol. 19, No. 4, 2015,

pp. 355–369. http://dx.doi.org/10.1080/15472450.2014.941750.

13. Liu, H., A Tok, and G. S. Ritchie. Development of a Real-Time On-Road Emissions Estimation and Monitoring System. *Proc., 14th International IEEE Conference on Intelligent Transportation Systems*, Washington, D.C., 2011. http://ieeexplore.ieee.org/stamp/stamp.jsp?arnumber=6083006. Accessed May 22, 2017.
14. Oh, C., and G. S. Ritchie. Recognizing Vehicle Classification Information from Blade Sensor Signature. *Pattern Recognition Letters*, Vol. 28, No. 9, 2007, pp. 1041–1049. http://dx.doi.org/10.1016/j.patrec.2007.01.010.
15. Coifman, B. Using Dual Loop Speed Traps to Identify Detector Errors. *Transportation Research Record: Journal of Transportation Research Board*, 1999. 1683: 47–58. https://doi.org/10.3141/1683-07.
16. Zhang, G., Y. Wang, and H. Wei. Artificial Neural Network Method for Length-Based Vehicle Classification Using Single-Loop Outputs. *Transportation Research Record: Journal of the Transportation Research Board*, 2006. 1945: 100–108.
17. Coifman, B., and S. Kim. Speed Estimation and Length Based Vehicle Classification from Freeway Single-Loop Detectors. *Transportation Research Part C- Emerging Technologies*, Vol. 17, 2009, pp. 349–364. https://doi.org/10.1016/j.trc.2009.01.004.
18. Weinblatt, H., E. Minge, and S. Peterson. Length-Based Vehicle Classification Schemes and Length Bin Boundaries. *Transportation Research Record: Journal of the Transportation Research Board*, 2013. 2339: 19–29.
19. Liu, H. X., and J. Sun. Length-Based Vehicle Classification Using Event-Based Loop Detector Data. *Transportation Research Part C- Emerging Technologies*, Vol. 38, 2014, pp. 156–166. https://doi.org/10.1016/j.trc.2013.11.010.
20. Gardner, D. *Axle Correction Factors*. Ohio Department of Transportation, Columbus, OH, 2015. https://www.dot.state.oh.us/Divisi ons/Planning/TechServ/traffic/Reports/AxlCorrFctrs/AxleCorFC15.pdf. Accessed May 12, 2017.
21. Marti, M., R. Kuehl, and S. Peterson. *Traffic Data Collection Processes Study*. Publication MN/RC – 2014RIC51A. SRF Consulting Group, Minnesota Department of Transportation, Saint Paul, MN, 2014.

The Standing Committee on Highway Traffic Monitoring (ABJ35) peer-reviewed this paper (18-00324).

Article

TRR
JOURNAL OF THE TRANSPORTATION RESEARCH BOARD

Transportation Research Record
2018, Vol. 2672(43) 122–133

DOI: 10.1177/0361198118787364
journals.sagepub.com/home/trr
SAGE

Application of Trajectory Data for Investigating Vehicle Behavior in Mixed Traffic Environment

Narayana Raju[1], Pallav Kumar[1], Aayush Jain[1], Shriniwas S. Arkatkar[1], and Gaurang Joshi[1]

Abstract

The research work reported here investigates driving behavior under mixed traffic conditions on high-speed, multilane highways. With the involvement of multiple vehicle classes, high-resolution trajectory data is necessary for exploring vehicle-following, lateral movement, and seeping behavior under varying traffic flow states. An access-controlled, mid-block road section was selected for video data collection under varying traffic flow conditions. Using a semi-automated image processing tool, vehicular trajectory data was developed for three different traffic states. Micro-level behavior such as lateral placement of vehicles as a function of speed, instant responses, vehicle-following behavior, and hysteresis phenomenon were evaluated under different traffic flow states. It was found that lane-wise behavior degraded with increase in traffic volume and vehicles showed a propensity to move towards the median at low flow and towards the curb-side at moderate and heavy flows. Further, vehicle-following behavior was also investigated and it was found that with increase in flow level, vehicles are more inclined to mimic the leader vehicle's behavior. In addition to following time, perceiving time of subject vehicle for different leading vehicles was also evaluated for different vehicle classes. From the analysis, it was inferred that smaller vehicles are switching their leader vehicles more often to escape from delay, resulting in less following and perceiving time and aggressive gap acceptance. The present research work reveals the need for high-quality, micro-level data for calibrating driving behavior models under mixed traffic conditions.

Driver behavior on road sections, is one of the most intricate phenomena to study, particularly under mixed traffic environment. This is also interconnected with the competence of the road section and the traffic flow characteristics of the road section at macroscopic as well as microscopic level. To capture the driver's behavior and compute its nature, many research studies are conducted, and outcomes have resulted in different driving behavior models. The driving behavior models include mainly car-following, lane-changing and lateral-behavior models particularly for mixed traffic non-lane-based conditions. Under mixed traffic conditions, the following behavior is vehicle-class-based and lateral interaction also depends hugely on type of subject vehicle and the number and type of vehicles surrounding it. Other than following, lane changing, lateral movement, filtering or seeping, oblique following, and staggered following, there can be group behavior, and the dynamics are extremely difficult to understand without the availability of high-quality trajectory data.

Driving behavior models are generally developed expressing certain phenomena over the road sections. On this basis, Pipes (*1*) framed a concept and related vehicular-following behavior to law of separation and specified some prescribed following distance, based on the follower velocity. Gazis et al. (*2*) in their work formulated the acceleration of follower vehicle, which is directly proportional to the relative velocity among the vehicles. To explain the following behavior, Gipps (*3*) classified the response of following vehicles into two regimes, with an assumption that driver sets limits to his desired braking and acceleration rates. Bando et al. (*4*) presented a dynamic model based on equations of motions and coined a term legal-velocity, which is considered as a function of headway from the preceding vehicle. Newell (*5*) proposed a simple car-following model based on the time–space plots of vehicles. It was observed initially that researchers had attempted modelling the subject follower response to its leader stimulus, but most of the studies were limited to a certain homogeneous traffic condition. In the later stages, researchers had modified the well-known car-following models and evaluated their performance over varying traffic scenarios, Chakroborty et al. (*6*) in their work evaluated the

[1]Civil Engineering Department, Sardar Vallabhbhai National Institute of Technology, Surat, Gujarat, India

Corresponding Author:
Address correspondence to Shriniwas S. Arkatkar: sarkatkar@gmail.com

General Motors (GM) model and added fuzzy concept to model the following behavior. Rakha et al. (*7*) in his work calibrated different car-following models and developed macroscopic fundamental plots for the study section and validated with observed data. Further, with the advancement in technology and affordability of high-computational tools, in recent times, car-following models were programmed implicitly in the simulation tools and played a huge role in simulating the road networks for finding optimized solutions. On this basis, Aycin et al. (*8*) compared the performance of car-following models and concluded that models are giving similar outputs with different car-following approaches. Kesting et al. (*9*) highlighted the importance of vehicular trajectory for modeling the car-following phenomenon and proposed a methodology for calibrating car-following models. Toledo et al. (*10*) emphasized the importance of lane-changing behavior and classified lane-changing into mandatory and discretionary and modelled integrated lane-change behavior. Ben-Akiva et al. (*11*) modelled the lane-change models among vehicles and compared their performance with the existing lane-change models. From the literature, it may be inferred that most of the formulated driving behavior models relate to homogeneous lane-based traffic conditions.

Under heterogeneous non-lane-based traffic conditions, the interactions among the vehicles are much more complex in nature. Due to heterogeneity of traffic the driving behavior is substantially varying from homogeneous lane-based traffic behavior. Because of the driving behavior involved over the road sections, even the well-established, automated image processing tools used in homogeneous traffic conditions for trajectory development are failing drastically in developing vehicular trajectory data sets on these road sections. Due to the lack of vehicular trajectory sets, very few studies were reported explaining the driving behavior. In this direction, Arkatkar et al. (*12*) analyzed the effect of gradient on the performance of the vehicles and quantified its effect on macro-level characteristics of the road section using simulation technique. With the help of vehicular trajectory, Kanagaraj et al. (*13*) assessed the lateral behavior of vehicles and reported variation based on vehicle-type. Kanagaraj et al. (*14*) explained the merging behavior of vehicles in an Indian traffic context. To develop vehicular trajectories under mixed traffic and study their nature Liu et al. (*15*) presented an approach and proposed anchor-based calibration system with the help of GPS in developing vehicular trajectory sets. Even the authors Raju et al. (*16*) modelled the following behavior by means of trajectory data and evaluated the modelled behavior using simulation. To understand the performance of road sections, Da Silva et al. (*17*) applied a concept of network-delta to quantify the performance for a heterogeneous traffic network. But more often network-delta is an ITS-based approach in quantifying the performance of road networks. From the available literature, it was inferred that very few studies had been attempted in assessing the vehicular behavior, even in homogeneous traffic conditions. Most of the studies are dependent on NGSIM data (*18*) which was developed for testing the modelled driving behavior models. But, under heterogeneous traffic conditions, due to lack of trajectory data, the explicit behavior of vehicles under mixed traffic conditions is not reported by the earlier studies. With this motivation, this research work is focused firstly on developing high-quality trajectory data under mixed traffic conditions, involving different vehicle categories maneuvering and interacting in longitudinal as well as lateral directions. After, developing this high-resolution trajectory data, some of the driving patterns are captured and reported in this study.

Study Section

In this study, an uninterrupted, mid-block road section was selected on a multilane, high-speed road section in Mumbai, India to examine driving behavior. The roadway section is divided into ten lanes and has a lane marking every 3.5 m. The road section was selected in such a way that varying roadway and traffic characteristics are accounted for, so that results would have wide applicability. A video camera was installed at a vantage point on the study location to capture the continuous traffic flow and identify the vehicles along the sections distinctly for a trap length of 120 m. The data was collected for twelve hours to capture different traffic flow conditions prevailing at different time periods of the selected day. On this basis around twelve hours of video graphic data were collected over the study section and snap shots of the traffic flow at different flow levels were depicted in Figure 1*a*. From the surveys, it was found that around six categories of vehicles were observed over the study section, namely auto rickshaw (Auto), motorized two-wheeler (Bike), car, bus, truck, and light commercial vehicle (LCV).

Traffic-State Estimation

It was decided to initially identify traffic states and classify the whole macroscopic traffic data into an optimal number of clusters. This step would help in understanding differences in micro-level variables such as acceleration, deceleration, and vehicle-to-vehicle interaction for different traffic flow levels. Once the data is well-classified, the duration of traffic data extraction can be selected in such a way that all traffic states can be included to get a very comprehensive trajectory data using a semi-automatic tool.

Based on the videography data from the study locations, the macroscopic fundamental plots were developed for the study section and the characteristics were normalized in terms of passenger car unit (PCU) (*19*). From the analysis, the maximum flow rate observed of 12000 PCU/H is considered as the capacity of the roadway section. In the present study, to study the behavior of vehicles over different flow level, initially, traffic-state analysis was conducted taking

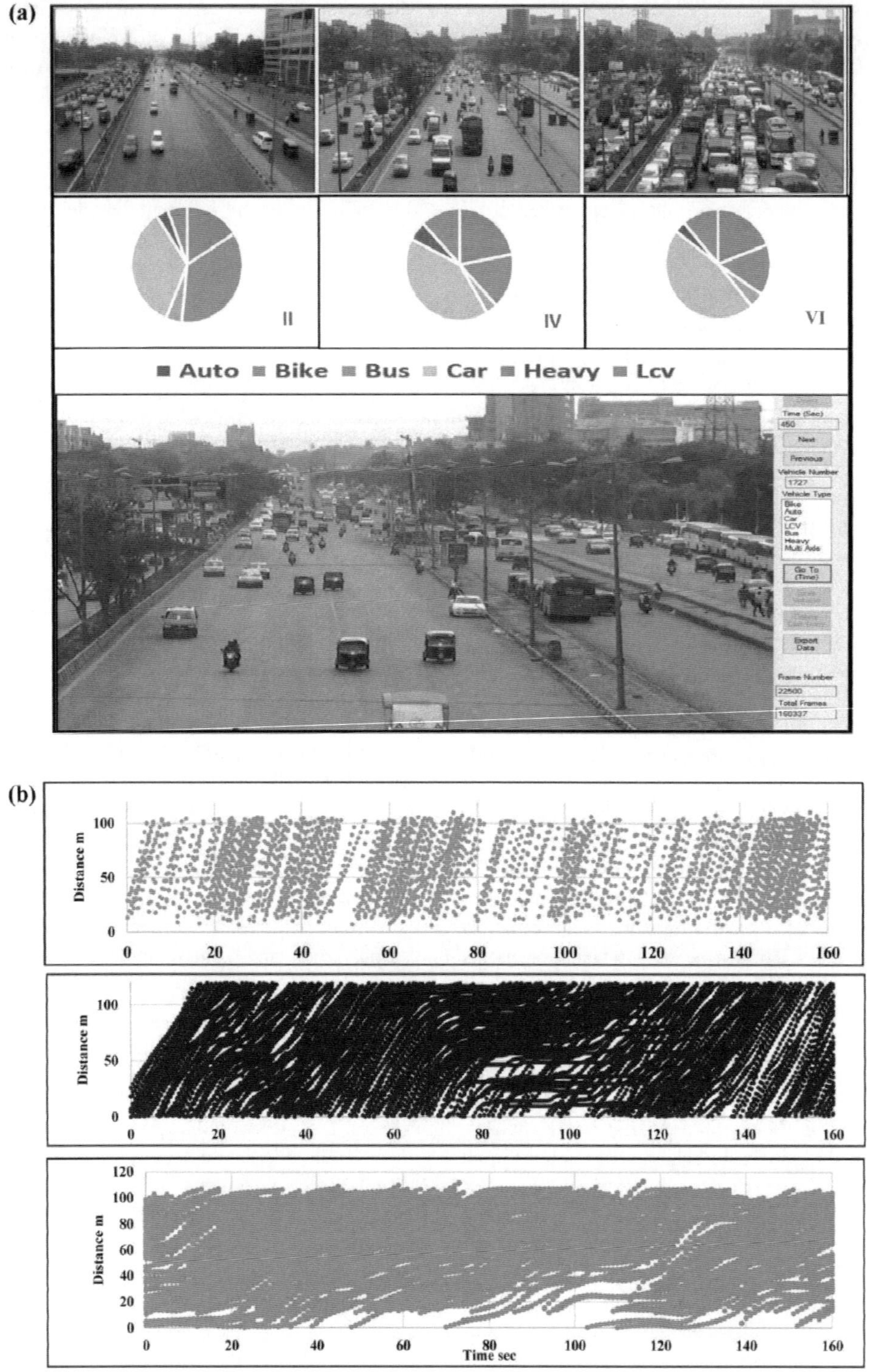

Figure 1. (*a*) Traffic flow, composition, trajectory data extractor interface and (*b*) time–space plots of vehicles at different volume levels.

Table 1. Traffic-State Clusters and Details of Extracted Trajectory Data Sets

Traffic-state cluster for study section

Traffic state	Flow (PCU/H)	Density (PCU/Km)	Area occupancy (%)	V/C ratio
I	<3200	<45	<0.02	<0.25
II	3200–5600	45–84	0.02–0.032	0.25–0047
III	5600–7800	84–130	0.032–0.048	0.47–0.65
IV	7800–10400	130–185	0.048–0.07	0.65–0.87
V	10400–12000	185–245	0.07–0.10	0.87–1.00
VI	>12000	>245	>0.10	–

Details of trajectory sets over the study section

Flow level	No. of vehicles observed	Duration of trajectory data	Area occupancy (%)	Traffic state
Low flow conditions	1079	10 minutes	0.025	II
Medium flow conditions	1902	15 minutes	0.061	IV
Heavy flow conditions	650	10 minutes	0.220	VI

area occupancy (*20*) as an independent variable. It was observed that data sets ranged from free-flow to capacity regime and a few points were found in congested regime also. The extracted data sets on speed, flow, and density (in terms of PCU) were used for the allocation of different ranges/classifications within it. To classify the thresholds of speed, flow, and density for different traffic states, k-mean clustering technique (*21*) was applied.

For the present study five clusters were determined to define traffic state to comply with international standards of uniformity (*22*). The data are combined for least distance from the closet cluster center using the Euclidean distance measure (*23*). Once the cluster membership of each variable is defined, data points are plotted for speed-flow and flow-density curves to define thresholds of traffic state by classified membership. The results of the traffic-state thresholds are given in Table 1, for the study section.

Trajectory Data Development

From the literature review, it was inferred that vehicular trajectory data is one of the high-quality, micro-level data sources, using which interactions among vehicles can be studied. Considering this, the present research work attempted to develop vehicular trajectory data under prevailing mixed non-lane-based traffic conditions. From the preliminary analysis, it was found that automated trajectory development tools are not that efficient for trajectory development under non-lane-based, mixed traffic conditions. Hence, traffic data extractor (*24*), a semi-automated tool, was used in developing trajectory data, in which the vehicle position was tracked by means of the mouse pointer with update interval of 0.5 sec with GUI as shown in Figure 1*a*. In this study, to analyse the vehicular behavior over different flow conditions, trajectory data was developed for three different varying flow conditions, ranging from free-flow (Traffic state I and II) to stop-and-go (Traffic state VI) conditions. For this purpose, around 3631 vehicles were tracked for a complete duration of 35 minutes. Similarly, to nullify the human errors, while extracting, trajectory smoothening was carried out with a 5-moving average (*25*).

Based on visual inspection, initially trajectory data were extracted by traffic-state (flow) level and were classified into trajectory sets of low, medium and heavy flows for a duration of 10, 15 and 10 mins, respectively. To classify the trajectory sets in a more scientific way, area occupancy of vehicles was evaluated over the duration as reported in Table 1. This analysis is important to report, based on the level the road network is serving. On this basis, the datasets were classified into II, IV and VI, as defined in Table 1. Similarly, the vehicular composition over these flow levels is also given in Figure 1*a* and the developed high-quality time–space plots of individual vehicles are shown in Figure 1*b*.

Micro-Level Analysis

Micro-level interaction of vehicles is one of the complicated phenomena under heterogeneous non-lane-based traffic. This study is mainly focused on vehicle behavior over different traffic flow conditions. In connection with this, the lateral placement of vehicles, instantaneous responses in terms of velocities, and following behavior of different vehicle types, are analyzed over different flow levels in this study and explained in the following sections.

Lateral Placement of Vehicles

Using trajectory data, the lateral behavior was investigated over the different flow levels (traffic states), for which frequency distributions were plotted across the entire road-width

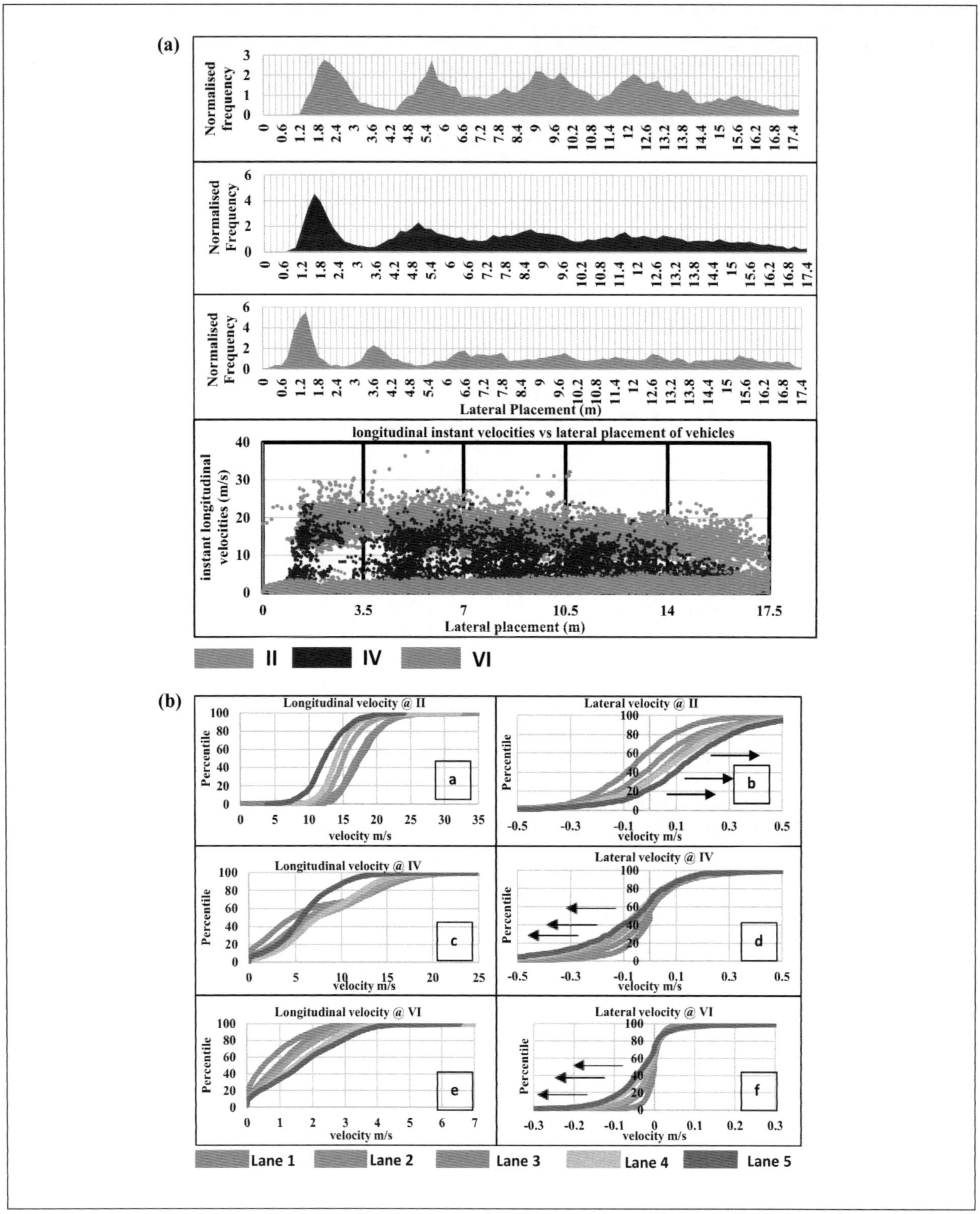

Figure 2. Micro-level traffic flow characteristics under different traffic state, showing (*a*) lateral placement and instant velocities of vehicles over different volume levels and (*b*) instant velocities of vehicles over different flow levels.

(in which origin was taken from the median side, with lane-1 at median-side, followed by other lanes). To compare the results, the frequencies were normalized to 100%. It was observed that in traffic state II, the vehicles are complying with lane-wise movement reasonably and moving in the specified lanes over the road section, which resulted in five crests in the frequency distribution, which can be observed in Figure 2*a*. It can be also noted that the vehicles are maintaining a shy-gap of around 0.6 to 0.8 m from the median. In the case of traffic state IV, the lane discipline was only observed in the median-side lane and its adjacent lane, whereas the other three lanes were operating in a non-lane-based sense. From this it may be inferred that heterogeneity of the traffic plays a strong role in degrading the compliance of adhering to lane-wise movements. This may be attributed to the dynamic and static nature of vehicular characteristics. It was also observed that shy-gap from median is decreased from 0.6 to 0.3 m (as traffic moves from traffic state II to traffic state IV). In the case of traffic state VI, a completely different kind of behavior of vehicles was observed, in which the smaller size vehicles (mainly auto: motorized three-wheelers and bikes; motorized two-wheelers) are acting as small particles passing through the porous medium created by cars and trucks. This driving behavior is called seepage or infiltration movement of smaller vehicles. Because they are small, these vehicles always place themselves in the lanes which are near the curb side to have a fast movement, and when compared with other vehicles resulted in non-lane movement on those lanes. On the other hand, the vehicles in the median side are arrested laterally and resulted in lane-wise behavior only on that single lane which can be visualized in Figure 2*a*.

Like lateral placement analysis, the instantaneous responses in terms of longitudinal and lateral velocities were investigated for all vehicles at aggregated level over the entire road width at different flow levels using exhaustive trajectory data. As shown in Figure 2*a*, instant longitudinal velocities were compared over the road section. From the analysis, it was found that for traffic state II, the vehicles on the median-side lane are moving at higher velocities, with a downward trend observed among the longitudinal velocities as they move towards the curb-side direction. Whereas for traffic state IV, no such peculiar trend was observed among the longitudinal velocities except variation in range over the lanes. Even shy-gaps were observed among these flow levels. On the other hand, a completely different scenario was observed for traffic state VI (congested), when compared with traffic state IV, in which the vehicles on the median-side lane have lesser longitudinal velocities, and an upward trend was observed as they move towards the curb side. To understand the behavior in depth, lane-wise instant longitudinal and lateral velocity percentile distributions were plotted over different traffic states as shown in Figure 2*b* (negative sign indicates movement towards curb side). The phenomenon is explained below:

Traffic State II: At this flow level, it was observed that the vehicles on the median-side lanes (i.e., lane 1 and lane 2) are operating at higher longitudinal velocities and velocities tend to decrease as they move towards the curb-side lanes (i.e., lane 4 and lane 5). Whereas, in the case of lateral velocities, the lanes which are close to the curb side have higher lateral velocities and they tend to decrease as they move towards the median-side. From this, it may be realized that the median-side lanes have lane-based behavior with minimum lateral movement and possess higher longitudinal velocities and vice versa with the curb-side lanes. From the lateral speed distributions, it may be observed that they are skewed towards the right side (positive side) as shown in Figure 2*b*(b), which shows that at low volume level, the vehicles are showing a propensity to move towards the median-side to have faster movement.

Traffic State IV: Unlike the previous scenario, there is no such peculiar behavior observed in longitudinal velocities of vehicles over the lanes, other than lane 5 operating at less longitudinal velocities as shown in Figure 2*b*(c). But in the case of lateral velocities, from the frequency distributions, it can be observed that they are more skewed towards the left side (negative side) as shown in the Figure 2*b*(d). Unlike in traffic state II, the vehicles are showing a tendency to move away from the median, to have faster movement over the road section.

Traffic State VI: In this flow level, it was observed that vehicles are moving in stop-and-go conditions. Because of very high density, the overall instantaneous velocities are reduced across all the lanes. It was found that the median-side lanes (lane 1 and lane 2) are occupied by cars; because of their size, they are just following one after another in these lanes. Whereas, the smaller vehicles, by virtue of size, are able to find the gaps in the traffic stream, which are more likely to be on the curb-side lanes. Due to this phenomenon the instantaneous longitudinal velocities are slightly higher on the curb-side lanes (lane 4 and lane 5), as shown in the Figure 2*b*(e). Similar phenomenon is also replicated in lateral velocities, in which the percentile distribution was more skewed towards the left side (negative side) and shows the vehicles are moving toward the curb side as shown in in Figure 2*b*(f).

From the trends observed for variation in instantaneous velocities, overall, it may be inferred that, at low flow level, on multi-lane high-speed urban roads, even under mixed traffic conditions, vehicles may follow reasonable lane-discipline (quasi-type). As the volume increases, the movement of the vehicles is affected resulting in a slower movement, thereby heterogeneity involved in traffic deviates from the lane-based behavior. In this phase, it is also observed that the small-size vehicles act as small particles and there can be seepage of small vehicles among the group of other larger vehicles. From the analysis, it was inferred that lateral behavior is more predominant in modeling the behavior of drivers under mixed non-lane-based traffic conditions.

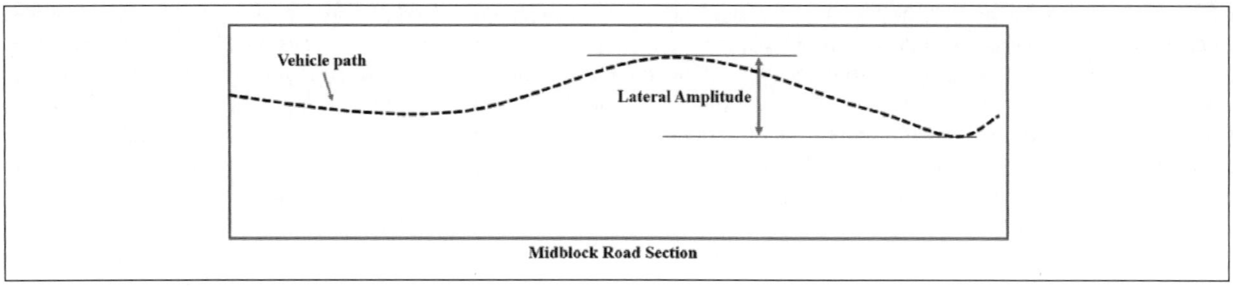

Figure 3. Schematic explanation of lateral amplitude.

Table 2. Lateral Amplitude of Vehicles Over the Study Section

	Traffic state II			Traffic state IV			Traffic state VI		
Vehicle type	Mean (m)	SD (m)	Range (m)	Mean (m)	SD (m)	Range (m)	Mean (m)	SD (m)	Range (m)
Auto	0.81	0.57	3.18	1.45	0.79	4.29	1.33	0.95	5.05
Bike	0.89	0.67	4.73	1.46	1.12	6.39	1.72	1.23	4.89
Bus	0.73	0.57	2.57	1.09	0.62	2.44	1.76	1.04	4.90
Car	0.79	0.75	4.34	1.01	0.94	6.49	1.13	1.10	9.59
Truck	0.49	0.22	1.18	0.93	0.72	2.89	1.66	1.03	4.87
LCV	0.65	0.53	2.82	0.90	0.63	3.32	1.80	1.22	6.55

Lateral Amplitude

To investigate the lateral behavior of vehicles in a deeper sense, lateral amplitude was computed for each vehicle category over different flow conditions (traffic states). Lateral amplitude is the lateral weaving of vehicles, which is the difference in maximum and minimum lateral positions over the study section, as shown in Figure 3. On this basis, mean, standard deviation (SD), and range, were calculated for each vehicle category, reported in Table 2. From the analysis, it may be inferred that as the flow over the road section increases the lateral amplitude values tends to increase for all the vehicles. On the other hand, the smaller vehicles (Auto and Bike) tend to have a more weaving nature, when compared with other vehicles for all flow levels. From this it can be inferred that mainly small vehicles are responsible for this non-lane-based movement of heavy vehicles at low flow levels.

Following Behavior

From the review of available literature, it is a well-known fact that most of the studies under these traffic conditions are concentered on modelling the car-following models to predict driver's behavior in following phase. But, the internal behavior among the following pairs are not explicitly studied under varying traffic flow levels, particularly under mixed traffic conditions. On this basis, the analysis was performed, and interesting inferences are drawn, as given in the next section.

Hysteresis Analysis

From the review of reported formulated car-following models, it is observed that the response of the follower (vehicle) is modelled using relative distances, free velocities, viewing angle, etc. The modelled responses can be accelerations/velocities of the follower. On this basis, Wiedemann (*26*) framed a relationship between relative distance with relative velocity (follower minus leader vehicle velocities) and explained different thresholds of following and proposed two psychophysical following-behavior models (Wiedemann74 and Wiedemann99). In this research work, to examine driving behavior under mixed traffic conditions in India, the logic of Wiedemann is used, wherein the relative distance versus relative velocity plots are made for successive vehicles on a given lane (considering width as 3.5 m). From this, it was observed that there is kind of hysteresis phenomenon (representation of the following behavior) observed among the vehicles, as shown in Figure 4. It shows nine different combinations of hysteresis among vehicles, which are presented, as examples, in which hysteresis plots 1, 4, 6, and 7 are showing close-up behavior, hysteresis plots-2, 3, and 8 depict good following interaction, whereas hysteresis plot-5 shows mixed following behavior and hysteresis plot-9 shows, kind of passing behavior.

On this basis, the relative distance versus relative velocity plots (hysteresis plots) were aggregated based on the follower vehicle category for each flow level. To check the behavior of vehicles at different flow levels, these aggregated hysteresis plots were then compared across following

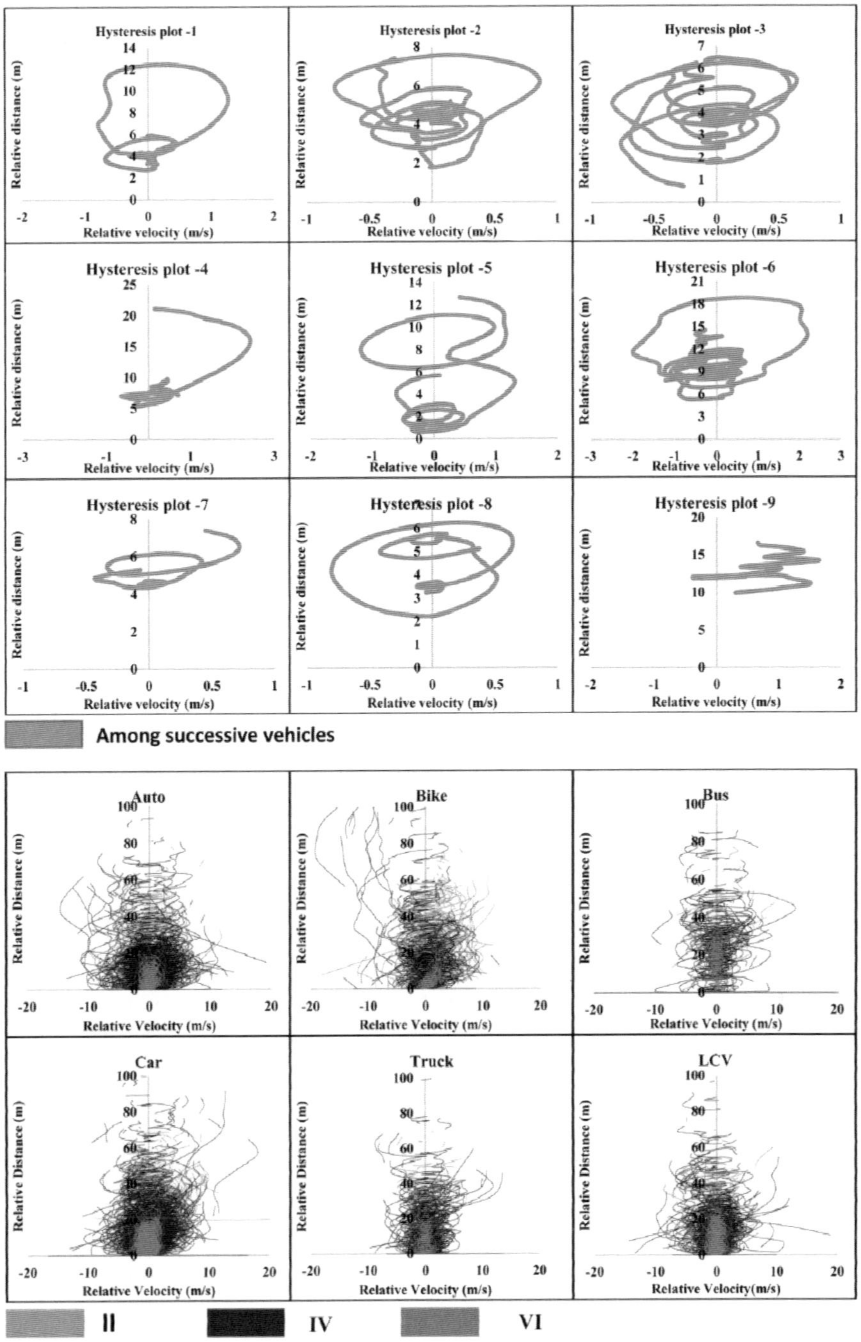

Figure 4. Hysteresis phenomenon of vehicles along with aggregated hysteresis plots of the vehicles.

vehicle category as reported in Figure 4. On this basis, to understand the behavior, descriptive statistics such as mean, standard deviation, range, and percentile ranges were computed for relative velocities and relative distances for each following vehicle category as reported in Table 3. From the analysis, it was inferred that for traffic state II, where the vehicles are in free-flow regime, it was observed that the interactions related to following behavior were not very effective and showed passive behavior with the leading vehicle, as this behavior is predominant in bikes and resulted in higher relative velocities when compared with other vehicle categories. For traffic state IV, vehicle-following behavior

Table 3. Descriptive Analysis on Aggregated Hysteresis of Following Pairs

Relative velocities analysis from the hysteresis plots

Traffic state	Units (m/s)	Auto	Bike	Bus	Car	Truck	LCV
II	Min	−10.04	−18.16	−7.32	−8.96	−5.82	−8.53
	Max	11.51	25.91	19.15	21.60	7.90	9.25
	Mean	−0.36	0.90	1.27	0.99	0.61	−0.27
	SD	2.43	2.90	3.08	2.44	2.57	2.65
	Range	21.55	44.06	26.47	30.56	13.72	17.79
IV	Min	−14.42	−19.02	−20.85	−16.68	−9.75	−14.48
	Max	19.06	12.77	20.48	18.14	13.83	18.92
	Mean	0.13	−0.17	0.13	0.07	0.15	0.03
	SD	2.38	3.21	2.63	1.96	2.31	2.26
	Range	33.48	31.79	41.34	34.82	23.58	33.40
VI	Min	−2.77	−2.17	−3.66	−6.68	−5.12	−5.26
	Max	2.97	3.95	3.03	7.41	3.11	6.89
	Mean	0.02	0.55	−0.05	−0.07	−0.11	0.00
	SD	0.60	0.91	0.75	0.62	0.85	0.67
	Range	5.74	6.12	6.69	14.09	8.23	12.15

Relative distance analysis from the hysteresis plots

Traffic state	Units (m)	Auto	Bike	Bus	Car	Truck	LCV
II	Range	78.4	83.4	80.5	80.5	83.1	82.3
	25th percentile	11.8	10.2	21.5	13.9	23.9	16.8
	50th percentile	20.2	18.1	29.3	20.8	32.6	23.8
	75th percentile	31.9	31.2	40.7	31.2	49.3	38.0
IV	Range	118.9	115.0	86.0	123.7	107.9	104.4
	25th percentile	4.2	5.1	10.2	5.2	2.3	2.5
	50th percentile	6.7	9.6	18.2	8.7	9.5	8.9
	75th percentile	13.1	22.3	32.1	14.5	18.7	16.2
VI	Range	48.8	38.7	50.0	55.2	41.4	53.0
	25th percentile	2.4	2.2	12.8	3.0	6.2	3.7
	50th percentile	4.2	3.8	18.7	4.7	10.8	8.4
	75th percentile	6.8	6.5	24.6	7.3	16.5	14.5

Following time of vehicles over the study section

Traffic state	Vehicle type	No. of samples	Mean (sec)	SD (sec)	Range (sec)
II	Auto	212	2.50	1.66	6.5
	Bike	466	2.44	1.62	7
	Bus	44	2.07	1.35	5.5
	Car	406	2.24	1.58	10.5
	Truck	42	1.74	1.47	5.5
	LCV	71	2.18	1.53	6.5
IV	Auto	840	7.32	8.56	56
	Bike	484	3.94	3.57	46.5
	Bus	125	6.96	7.07	28
	Car	1274	8.46	10.97	73.5
	Truck	212	10.68	12.40	55
	LCV	414	7.72	9.80	67.5
VI	Auto	409	12.26	15.06	82
	Bike	304	4.78	5.69	40
	Bus	104	18.65	18.99	78
	Car	1184	16.44	23.67	187
	Truck	112	15.64	16.33	66
	LCV	376	15.06	19.25	172

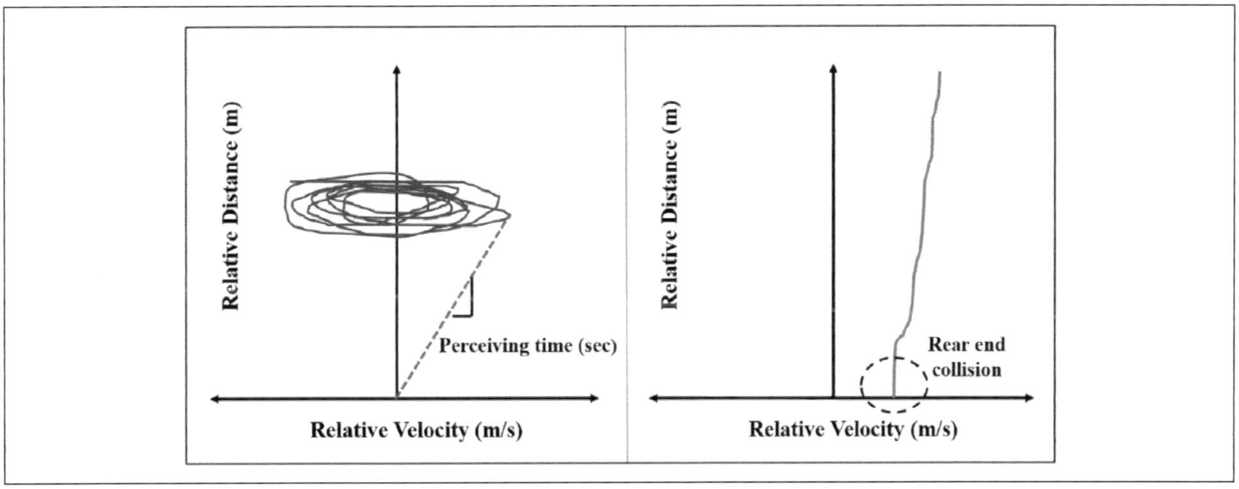

Figure 5. Schematic diagram explaining the perceiving time.

related to closing and giving up between vehicular pairs were observed among the vehicles, resulting in higher relative velocities than traffic state II in all vehicle categories. Due to this phenomenon higher relative velocity and relative distance ranges were observed in these flow levels. Whereas, in traffic state VI, the vehicles are experiencing a congestion state. In this phase the freedom related to lateral movement for the vehicles is very limited, and as a result, vehicles are just following the vehicles ahead. Due to this, a good amount of following-behavior pattern was observed among the vehicles. Even the relative distances and relative velocities are less, when compared with other flow levels. From the analysis, it was observed that with increase in vehicular flow the relative distances were becoming lesser for all kinds of vehicles. In the case of relative velocities, it was concluded that vehicles show aggressive behavior as subject to leader vehicles at traffic states II and IV. In congested state (traffic state VI), the vehicles are almost mimicking the leader vehicle's behavior.

Following Time Analysis

To have a clear-cut opinion on the following behavior of vehicles, other than hysteresis analysis, following time of vehicles had been taken up to observe the driving behavior. On this basis, the following time of vehicles were analyzed from the trajectory data sets, in which the amount of time spent by the vehicle behind another vehicle on the same lane had been evaluated for each vehicle by means of an algorithm. Based on the vehicle category the descriptive statistics such as mean, standard deviation, range along with vehicle pairs were computed for each vehicle category at different flow levels as reported in Table 3. From the analysis, it was clearly observed that small vehicles have extra freedom in switching their positions laterally over the road section, whereas heavy vehicles, because of their size, do not have significant lateral movement. Due to this, they have a high following time compared with smaller vehicle categories. It was inferred that under heterogeneous traffic conditions, mainly because of smaller vehicles, the lane discipline is affected and results in a complex driving behavior which is significantly different from homogeneous lane-based traffic behavior.

Perceiving Time Analysis

To understand the follower-vehicle perceiving time with leader vehicle, the perceiving nature of the subject follower-vehicle behavior is also examined. When a subject follower is moving closer to the leader vehicle, at a given distance, the following-subject vehicle perceives that he/she is moving closer to the leader and tends to match the leader vehicle's speed. In this case, the time available for the following vehicle to mimic the leader, to avoid a rear-end collision, is coined as perceiving time in this study. The lesser the perceiving time, the greater the chances of rear-end collision among the vehicles, as shown in the self-explanatory plots in Figure 5. In this context, to compute the perceiving time, hysteresis plots were examined in detail. It was observed that at the time of perceiving, the follower vehicle just tends to start decreasing its velocity to match the leader velocity. In this phase, the difference in velocity will be maximum (positive side) among the vehicles. With the help of that, the perceiving time available for the driver is computed as the slope of the line joining the point having maximum relative velocity on positive side.

On this basis, the perceiving time was computed for all successive vehicles in each lane and they were aggregated based on the subject vehicle category. Similarly, the mean and standard deviation were computed based on vehicle

Table 4. Perceiving Time of the Vehicles for a Given Follower Vehicle Type

	Traffic state II		Traffic state IV		Traffic state VI	
Vehicle type	Mean (sec)	SD (sec)	Mean (sec)	SD (sec)	Mean (sec)	SD (sec)
Auto	38.5	128.2	21.5	126.1	21.9	129.9
Bike	19.0	46.8	15.9	60.5	4.2	7.4
Bus	43.9	107.4	37.5	138.3	22.6	15.0
Car	31.5	120.5	21.5	119.4	10.3	15.1
Truck	32.8	34.9	15.0	31.7	16.9	23.9
LCV	81.8	227.5	15.9	31.5	10.5	11.4

category, at different flow levels, as reported in Table 4. From the statistical analysis, it was observed that perceiving time decreases as the flow level increases. This shows that vehicles are moving closer and having more risk of a rear-end collision with flow level. Whereas, at a given flow level, the smaller vehicles, mainly bikes, have less perceiving time compared to heavy vehicles, which shows the aggressiveness of smaller vehicles.

Important Findings

Based on the framework adopted in this research, the behavior of vehicles in the traffic stream had provided a few good insights, indicating need of high-resolution trajectory data. On the basis of this, the following are the inferences made:

- It is quite evident that driving behavior in heterogeneous conditions is a more complex phenomenon than in homogeneous conditions. To express the mixed traffic behavior, a single, two-dimensional vehicle-following model will not replicate the real field conditions. This is mainly due to complexity observed in driving patterns, different maneuvers, including lateral movement of vehicles, particularly of smaller vehicles, as indicated in the present study.
- It was found that smaller vehicles show more lateral-movement freedom compared with heavy vehicles and their presence may result in deviation of lane-wise movement, particularly whenever they experience delay. Similarly, those vehicles can be potentially mainly responsible for degrading the traffic safety on the sections.
- When compared to homogeneous, lane-based traffic conditions, the following behavior of vehicles can be greatly influenced by lateral behavior under heterogeneous non-lane-based traffic. Based on the studies of this kind, a new following-behavior model inclusive of lateral behavior can be coded in simulation tools for heterogeneous traffic conditions.
- From the findings of the study, small vehicles are seeping out to escape the delay under heavy flow conditions. Using this logic, the behavioral parameter in simulation models can be coded. In a similar way, non-lane-based coding of road networks, following behavior inclusive of lateral behavior, flocking nature, modelling impedance for side friction with volume levels, and lateral weaving of vehicles with vehicle size etc. must be modelled in microsimulation tools for representing heterogeneous traffic conditions

Conclusion

From the study, it is quite clear that heterogeneity in traffic plays a huge role in influencing the traffic behavior. Due to lack of high resolution trajectory data, unlike homogenous traffic conditions, very few studies had been carried out in this direction for assessing driving behavior. It was clearly observed that lateral behavior of vehicles plays a vital role in the driving behavior and its patterns, which is not very predominant under homogeneous traffic conditions with perfect lane discipline. Despite this, the established car-following models for mixed conditions are not fully successful in explaining the behavior of vehicles in mixed traffic. With opportunities for autonomous vehicle traffic environments in the near future, there is a huge research gap in this direction to code driving behavior. In that case the present approach of using high-quality trajectory data for identifying varying driving patterns is a way forward to model traffic effectively in mixed traffic conditions.

Author Contributions

The authors confirm contribution to the paper as follows: study conception and design: Arkatkar, Joshi; data collection: Kumar, Raju; analysis and interpretation of results: Raju, Kumar, Jain; draft manuscript preparation: Raju, Arkatkar. All authors reviewed the results and approved the final version of the manuscript.

References

1. Pipes, L. A. An Operational Analysis of Traffic Dynamics. *Journal of Applied Physics*, Vol. 24, No. 3, 1953, pp. 274–281. https://doi.org/10.1063/1.1721265.
2. Gazis, D. C., R. Herman, and R. B. Potts. Car-Following Theory of Steady-State Traffic Flow. *Operations Research*, Vol. 7, No. 4, 1959, pp. 499–505. https://doi.org/10.1287/opre.7.4.499.

3. Gipps, P. G. A Behavioural Car-Following Model for Computer Simulation. *Transportation Research Part B: Methodological*, Vol. 15, No. 2, 1981, pp. 105–111. https://doi.org/10.1016/0191-2615(81)90037-0.
4. Bando, M., K. Hasebe, A. Nakayama, A. Shibata, and Y. Sugiyama. Dynamical Model of Traffic Congestion and Numerical Simulation. *Physical Review E*, Vol. 51, No. 2, 1995, pp. 1035–1042. https://doi.org/10.1103/PhysRevE.51.1035.
5. Newell, G. F. A Simplified Car-Following Theory: A Lower Order Model. *Transportation Research Part B: Methodological*, Vol. 36, No. 3, 2002, pp. 195–205. https://doi.org/10.1016/S0191-2615(00)00044-8.
6. Chakroborty, P., and S. Kikuchi. Evaluation of the General Motors Based Car-Following Models and a Proposed Fuzzy Inference Model. *Transportation Research Part C: Emerging Technologies*, Vol. 7, No. 4, 1999, pp. 209–235. https://doi.org/10.1016/S0968-090X(99)00020-0.
7. Rakha, H., and B. Crowther. Comparison of Greenshields, Pipes, and Van Aerde Car-Following and Traffic Stream Models. *Transportation Research Record: Journal of the Transportation Research Board*, 2002. 1802: 248–262.
8. Aycin, M., and R. Benekohal. Comparison of Car-Following Models for Simulation. *Transportation Research Record: Journal of the Transportation Research Record,* 1999. 1678: 116–127.
9. Kesting, A., and M. Treiber. Calibrating Car-Following Models by Using Trajectory Data: Methodological Study. *Transportation Research Record: Journal of the Transportation Research Board*, 2008. 2088: 148–156.
10. Toledo, T., H. N. Koutsopoulos, and M. Ben-Akiva. Modeling Integrated Lane-Changing Behavior. *Transportation Research Record: Journal of the Transportation Research Board*, 2003. 1857: 30–38.
11. Ben-Akiva,M., C. Choudhury , and T. Toledo. Integrated Lane Changing Models *Transport Simulation: Beyond Traditional Approaches*, Chung E. and Dumont A-G., eds., CRC Press, pp. 61–74, 2009.
12. Arkatkar, S. S., and V. T. Arasan. Effect of Gradient and Its Length on Performance of Vehicles under Heterogeneous Traffic Conditions. *Journal of Transportation Engineering*, Vol. 136, 2010, pp. 1120–1136. https://doi.org/10.1061/(ASCE)TE.1943-5436.0000177.
13. Kanagaraj, V., G. Asaithambi, T. Toledo, and T. C. Lee. Trajectory Data and Flow Characteristics of Mixed Traffic. *Transportation Research Record: Journal of the Transportation Research Board*, 2015. 2491: 1–11.
14. Kanagaraj, V., K. K. Inivasan, and R. Sivanandan. Modeling Vehicular Merging Behavior Under Heterogeneous Traffic Conditions. *Transportation Research Record: Journal of the Transportation Research Board*, 2011. 2188: 140–147.
15. Liu, S., C. Luo, Q. Luo, L. M. Ni, and R. Krishnan. Calibrating Large Scale Vehicle Trajectory Data. *Proceedings of the 2012 IEEE 13th International Conference on Mobile Data Management* pp.222–231. 2012.
16. Raju, N., P. Kumar, A. Chepuri, S. S. Arkatkar, and G. Joshi. Calibration of Vehicle Following Models using Trajectory Data Under Heterogeneous Traffic Conditions. Presented at 96th Annual Meeting of the Transportation Research Board, Washington, D.C., 2017.
17. Da Silva, C. M., and W. Meira. Evaluating the Performance of Heterogeneous Vehicular Networks. *Proc., 2015 IEEE 82nd Vehicular Technology Conference (VTC2015-Fall)*, Boston, Mass., IEEE, New York, pp 1–5, 2016.
18. Punzo, V., M. T. Borzacchiello, and B. Ciuffo. On the Assessment of Vehicle Trajectory Data Accuracy and Application to the Next Generation SIMulation (NGSIM) Program Data. *Transportation Research Part C: Emerging Technologies*, Vol. 19, No. 6, 2011, pp. 1243–1262. https://doi.org/10.1016/j.trc.2010.12.007.
19. Kumar, P., S. S. Arkatkar, G. Joshi, and A. Dhamaniya. New Methodology for Estimating PCU on Multi Lane Urban Roads under Mixed Traffic Scenario Based on Area Occupancy. Presented at 96th Annual Meeting of the Transportation Research Board, Washington, D.C., 2017.
20. Mallikarjuna, C., and K. R. Rao. Area Occupancy Characteristics of Heterogeneous Traffic. *Transportmetrica*, Vol. 2, No. 3, 2006, pp. 223–236. https://doi.org/10.1080/18128600608685661.
21. Sangalli, L. M., P. Secchi, S. Vantini, and V. Vitelli. K-Mean Alignment for Curve Clustering. *Computational Statistics and Data Analysis*, Vol. 54, No. 5, 2010, pp. 1219–1233. https://doi.org/10.1016/j.csda.2009.12.008.
22. Brilon, W., and J. Lohoff. Speed-Flow Models for Freeways. *Procedia-Social and Behavioral Sciences*, Vol. 16, 2011, pp. 26–36.
23. Gower, J. C. Euclidean Distance Geometry. *Mathematical Scientist*, Vol. 7, No. 1, 1982, pp. 1–14.
24. Munigety, C., V. Vicraman and T. Mathew Semiautomated tool for extraction of microlevel traffic data from videographic survey, *Transportation Research Record: Journal of the Transportation Research Board*, 2443: 88–95.
25. Filters, M. A. Moving Average Filters. *Economics Letters*, Vol. 37, No. 4, 1991, pp. 277–284. https://doi.org/10.1016/0165-1765(91)90077-X.
26. Wiedemann, R. *Simulation Des Straßenverkehrsflusses*. Schriftenreihe des Instituts für Verkehrswesen, Universität Karlsruhe, 1974.

The Standing Committee on Highway Traffic Monitoring (ABJ35) peer-reviewed this paper (18-05165).

Article

TRR
JOURNAL OF THE TRANSPORTATION RESEARCH BOARD

Transportation Research Record
2018, Vol. 2672(43) 134–144

DOI: 10.1177/0361198118792338
journals.sagepub.com/home/trr
SAGE

Annual Average Nonmotorized Traffic Estimates from Manual Counts: Quantifying Error

Dylan Johnstone[1], Krista Nordback[2], and Sirisha Kothuri[3]

Abstract

Across the United States, jurisdictions are investing more in bicycle and pedestrian infrastructure, which can benefit from nonmotorized traffic volume data. The design of nonmotorized counting programs varies. Whereas some agencies use automated counters to collect continuous and short duration counts, the most common type of bicycle and pedestrian counting is manual counting either in the field or from video. The objective of this research is to identify the optimal times of day to conduct manual counts for the purposes of accurately estimating annual average daily nonmotorized traffic (AADNT). This study used continuous bicycle and pedestrian counts from six U.S. cities to estimate AADNT and analyze estimation errors for multiple short duration count scenarios. Using two permanent counters per factor group reduces error substantially (> 50%); afternoon counts seem to be best for reducing error (2:00 to 6:00 p.m.). Error on Sunday is often as good as, if not better than, Saturday, contrary to what others have found. Arlington has the lowest AADNT estimation error (mean absolute percentage error), probably because of better data quality and higher nonmotorized traffic volumes, and Mount Vernon, Washington has the highest. Average AADNT estimation errors for the studied short duration count scenarios range from 30% to 50%. Error is lower for the commute factor group, bicycle-only counts, scenarios in which more peak hours are counted, and when more than one permanent counter is available to estimate adjustment factors.

Annual average daily nonmotorized traffic (AADNT) estimates are foundational metrics that are useful for the planning, design, and engineering of bicycle and pedestrian infrastructure. Reliable bicycle and pedestrian traffic count data are needed to calculate accurate AADNT estimates. With limited funding available, bicycle and pedestrian counting programs often employ manual counters, typically volunteers, to count for a few hours per year at chosen locations. Despite previous research that documents 30% to 60% error for AADNT estimates from manual counts (*1*), the field continues to collect manual counts. Thus, optimizing and providing specific guidance for manual count programs would lead to higher quality data and better use of limited funds. This research seeks to answer questions related to when, how long, and how often to conduct manual counts to minimize this error. This study will contribute to the body of knowledge by investigating what times of day, days of the week, and times of year are associated with the lowest error for collecting manual counts.

This study used continuous counts from six U.S. cities to estimate AADNT and analyze estimation errors for multiple short duration count scenarios. Here the term AADNT is used to indicate bicycle-only, pedestrian-only, or bicycle-pedestrian-combined counts. The study cities were chosen because of the presence of a high number of permanent count sites. The study cities include Arlington, Virginia; Boulder, Colorado; Mount Vernon, Washington; Portland, Oregon; San Diego, California; and Seattle, Washington.

This research contributes to the growing literature on AADNT estimation. Unlike previous research, it investigates the accuracy of pedestrian-only, in addition to bicycle-only and bicycle-pedestrian-combined, counts. Additionally, this paper focuses on error estimates using less than 24 h of count data, whereas other studies assume that at least 24 h of count data are collected.

[1]Toole Design Group, Austin, TX
[2]Highway Safety Research Center, University of North Carolina, Chapel Hill, NC
[3]Department of Civil and Environmental Engineering, Portland State University, Portland, OR

Corresponding Author:
Address correspondence to Krista Nordback: nordback@hsrc.unc.edu

Background

For over a decade, the National Bicycle and Pedestrian Documentation Project (NBPDP) website has been the authoritative source on when and how to conduct manual counts (*2*). The FHWA Traffic Monitoring Guide (TMG) and other recent resources focus on automated counting strategies, but acknowledge that agencies may use short duration manual counts because of the perceived difficulty of using automated technologies and the desire to collect data on gender and helmet use (*3, 4*). The recently released NCHRP 797 reports that of those agencies surveyed who count pedestrians, 93% do so manually and of those who count bicycles, 87% do so manually (*5*). Although the NCHRP 797 provides valuable guidance on pedestrian and bicycle manual count programs, it is limited to reporting existing practice and does not investigate the optimal design of manual count programs.

The TMG suggests a minimum of 7 days (14 days preferred) for automated counts and 4 to 6 h for manual counts coinciding with the heaviest use (i.e., morning/evening for commute trips and midday for weekend/recreational trips) (*3*). When resources are available, 12-h short duration counts are preferred, which allow for time-of-day use profiles (i.e., travel patterns) to be calculated (*3*). These profiles assist in selecting the appropriate adjustment factors for estimation. If only manual counts are possible, the TMG encourages agencies to count at fewer locations for longer periods. "If non-motorized traffic levels are high and consistent from day-to-day, then shorter periods and/or fewer days may be considered. However, a longer-duration count period will be needed to determine how variable the non-motorized traffic is by time of day and day of week (DOW). Unfortunately, there is little quantitative guidance or consensus in this area, and ongoing research will improve future guidance" (*3*).

Research supports this guidance by demonstrating that average absolute percentage error (AAPE) also known as mean absolute percentage error (MAPE) in AADNT estimation decreases as count duration increases (*1, 6–10*). Previous research found 1 week of short duration counts is recommended for minimizing annual average daily bicyclist (AADB) estimation error, which requires automated counters (*1, 10, 11*).

The NBPDP recommends 2-h manual counts be taken on Tuesday, Wednesday, or Thursday during the traditional evening peak period (i.e., 5:00 to 7:00 p.m.) for weekday counts and on Saturday from noon to 2:00 p.m. for weekend counts. The NBPDP encourages counting during a designated week in mid-May and mid-September. The NBPDP also suggests counting during multi-hour periods up to three times at every location on sequential days and weeks, depending on the approximate use levels. Areas with high volumes (i.e., over 100 users per hour during midday periods) can typically be counted once for a weekday and a weekend day, except when there is unusual activity or nearby land use (i.e., special events, irregular activity at sports facilities, etc.), according to NBPDP (*2*).

Additionally, recent research has explored estimation error for counting bicycles on different days of the week and varying months or seasons. Common data collection methods suggest counting on Tuesday, Wednesday, and Thursday for weekdays, and Saturday for weekends. Research by Figliozzi et al. found Wednesday to produce the lowest error for weekdays and Saturday the lowest error for weekends (*9*). Past research found that counting during seasons with higher volumes and less variability results in better accuracy (e.g., May to October) (*1, 6, 7, 10, 12*). This period will vary with climate and location, but can be identified given 1 year of continuous count data. Further, the TMG suggests choosing months or seasons that are representative of average or typical use levels using continuous count data and recommends the NBPDP guidance for counting in the spring and fall months for most climates in the U.S. (*3*).

However, there is a lack of research on estimation error for time of day. Budowski found that using 2-h counts to estimate seasonal average daily bicyclists (SADB) in Winnipeg resulted in 12% to 23% MAPE, and the improvement in error for using 3- or 4-h counts did not warrant using additional resources (*8*). Budowski determined that the best time of day to conduct short duration counts is influenced by the observed traffic pattern (e.g., commuter, post-secondary commuter, recreational, mixed). For segments classified as commuter, the best time is from 3:00 to 5:00 p.m., 4:00 to 6:00 p.m. for recreational and mixed, and no analysis for post-secondary commuter (*8*). The study also confirmed that 1 week of counts is optimal for short duration counts (seven of eight study sites with < 10% MAPE). However, the study only examined bicycle-only data from multiuse paths in one city and studied seasonal rather than AADNT (*8*).

Recent research investigated factoring methods used for estimation and their associated accuracies. Day-of-year factoring provided more accurate estimates than the traditional AASHTO method and monthly factors (*10, 11*). Budowski found that a similar hour-of-year (disaggregate) factoring approach had similar lower error for SADB estimates based on hourly counts (*8*). However, in this study the researchers focused on the more traditional method similar to the TMG method because it may be simpler for agencies to implement.

Data

Continuous count data were obtained from Arlington, Boulder, Mount Vernon, Portland, San Diego, and

Table 1. Summary of Available Data

			Number of sites counted			
Community	Time period	Type of counters	Bike-only	Ped-only	Bike-ped-combined	Total number of sites
Arlington, Virginia	2012–2016	Passive infrared and inductive loop combination	18	11	NA	29
Boulder, Colorado	2002, 2004, 2007–2008, 2010–2012, 2016	Inductive loops	17	NA	NA	17
Mount Vernon, Washington	2009–2011	Passive infrared	NA	NA	6	6
Portland, Oregon	2009–2015	Active infrared, inductive loops, pneumatic tubes	9	8	15	32
San Diego, California	2013–2016	Passive infrared and inductive loop combination	33	13	NA	46
Seattle, Washington	2014–2016	Passive infrared and inductive loop combination, pneumatic tubes	9	7	NA	16
Total			86	39	21	146

Note: Bike = bicycle; Ped = pedestrian; NA = data not available.

Seattle from 2002 to 2016. To determine actual AADNT for a given year for a given site, at least one full 24 h set of count data representing each day of the week in every month was needed. From this data set, 146 locations had sufficient data for the analysis, totaling 1,461,604 hourly observations. A data summary is presented in Table 1.

The data were assessed for accuracy by the researchers. Suspicious data (i.e., consecutive zero counts for more than 48 h or volumes over 1,000 per hour) were manually identified and considered for exclusion from the analysis. Unusually low counts and zeroes perhaps resulting from temporary closures to the facility (e.g., holidays, snow events) or other events that could not be explained definitely were not removed. Some data were removed when a malfunction was evident (e.g., battery failure). Because of an oversight, partial count days were included in the analysis, but as there are few such days in the datasets, the results and findings were not significantly affected. Data were not removed for special events that increased bicycle and pedestrian traffic, although leaving these data in may result in higher error estimates. These events demonstrate the variability of nonmotorized traffic.

Methods

The analysis included four main parts described in the sections and in the flowchart shown in Figure 1: grouping sites, calculating nonmotorized traffic adjustment factors (e.g., hourly factors), AADNT estimation and error analysis, and a comparison of count scenarios.

Grouping Sites

Graphs of daily, weekly, and yearly temporal variations were plotted for each site. The sites were then grouped into three factor groups based on these patterns and a traffic distribution index proposed by Miranda-Moreno et al., the Average Morning/Midday Index (AMI) (*13*). The AMI is a ratio of morning to midday traffic. Average AMI is calculated using the following equation:

$$AMI = \frac{\sum_7^8 v_h}{\sum_{11}^{12} v_h} \quad (1)$$

where

AMI = Average Morning/Midday Index, and
v_h = Weekday average hourly count for hour, *h*.

The calculated AMI values were grouped using the following criteria: hourly noon activity (AMI < = 0.7), hourly multipurpose (0.7 < AMI < = 1.4), hourly commute (AMI > 1.4). Sites with noon activity have peak counts between the morning and evening peak hours. Sites demonstrate a commute pattern when the morning and evening peak hours are higher than the noon peak hour. Sites with multipurpose patterns generally have traffic patterns that mix both weekday commute and weekend recreational use patterns. The factor groups are

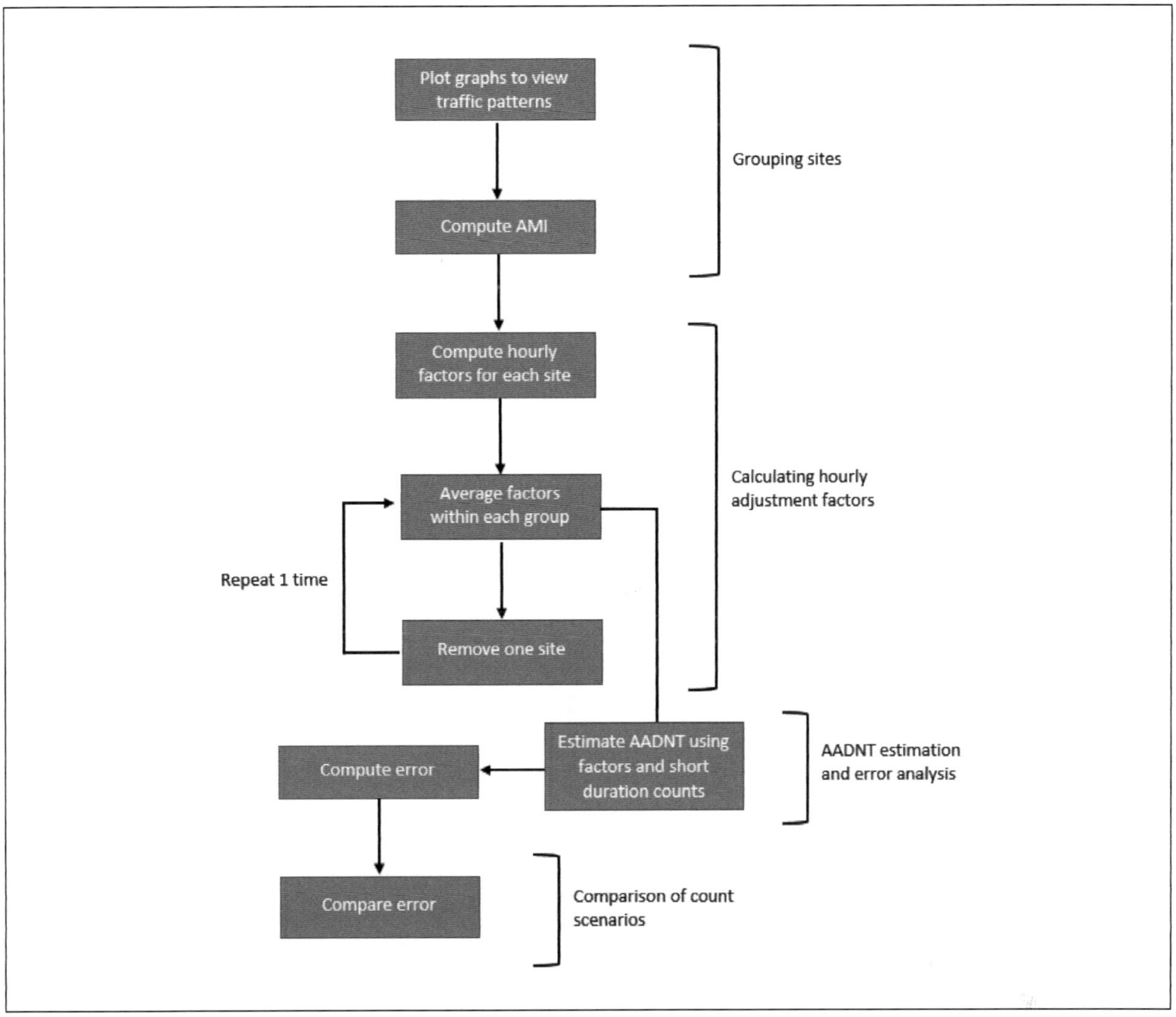

Figure 1. Flowchart showing steps followed in the analysis.

further separated by region and mode. A summary of this grouping is provided in Table 2.

Table 2 also includes ranges of AADNT values for each city and group. This table shows that all the cities have low volume sites (less than 200 AADNT) and all cities, except Mount Vernon, have high volume sites (greater than 600 AADNT). Mount Vernon is the smallest city, with fewer than 40,000 people, so it is not surprising that none of the count sites record above 200 AADNT.

Calculating Adjustment Factors

Hourly adjustment factors for every day of the week and month for a given year were computed. Because manual count programs collect less than 24 h of counts, an hourly factor is needed to adjust the hourly count up to an annual total. This hourly factor is calculated by dividing the average hourly traffic volume for that month and day of week by the AADNT.

$$\mathrm{AADNT}_{yks} = \frac{1}{12}\sum\nolimits_{m=1}^{12}\left[\frac{1}{7}\sum\nolimits_{j=7}^{1}\left[\frac{1}{n}\sum\nolimits_{i=1}^{n} V_{ijmyks}\right]\right] \tag{2}$$

where

AADNT_{yks} = actual annual average daily nonmotorized traffic for year, y, mode, k, and site, s;

V_{ijmyks} = total traffic volume for ith occurrence of the jth day of the week within the mth month, for year, y, for mode, k, and for site, s; and

n = the count of the jth day of the week during the mth month for which traffic volume is available (a number between 1 and 5).

Table 2. Summary of Factor Groups and AADNT

Community	Number of sites (*n*)	Hourly commute: Range of AADNT	*n*	Hourly multipurpose: Range of AADNT	*n*	Hourly noon activity: Range of AADNT	*n*
Arlington	29	67–1,756	22	22–3,588	7	NA	NA
Boulder	17	NA	NA	9–944	17	NA	NA
Portland	32	111–4,657	6	55–453	8	30–639	18
San Diego	46	128–289	3	18–9,216	22	19–15,354	21
Seattle	16	19–2,742	9	53–219	3	205–2,414	4
Mt. Vernon	6	NA	NA	NA	NA	14–163	6
Total	146	NA	40	NA	57	NA	49

Note: NA = missing data or inapplicable totals.

$$H_{hjmyks} = \frac{V_{hjmyks}}{\text{AADNT}_{yks}} \tag{3}$$

where

H_{hjmyks} = hourly factor for a given hour of the day, *h*, day of the week, *j*, month of the year, *m*, year, *y*, mode, *k*, and site, *s*;

V_{hjmyks} = average hourly nonmotorized traffic volume for a given hour of the day, *h*, day of the week, *j*, month of the year, *m*, year, *y*, mode, *k*, and site, *s*;

h = hour of the day (7:00 a.m.–7:00 p.m.);

j = day of the week;

m = month of the year;

y = year;

k = mode (bicycle, pedestrian, or bicycle-pedestrian-combined); and

s = site.

The average hourly traffic volume, V_{hjmyks}, is calculated using each hour from 7:00 a.m. to 7:00 p.m. for 7 days of the week for nonholidays in a given year to produce 1,008 factors per year for each site. For example, counts for a given site from the hour between 7:00 a.m. and 8:00 a.m. for all Tuesdays in May were averaged and divided by AADNT for that year and site. For this study, holidays are defined as federal holidays plus Thanksgiving week and December 25 to 31. Weeks are defined to start on Monday and end on Sunday.

Each hourly factor represents the average weather for a given hour of day, day of month, and month of year. Thus, this method inherently accounts for weather variations. Therefore, these hourly factors can be applied directly to estimate AADNT without applying daily or monthly factors.

Next, factors are computed by factor group (AMI groups) by averaging factors for sites in the group for each hour of the day, *h*, day of the week, *j*, month of the year, *m*, year, *y*, and mode, *k*. Then factors are recomputed after removing the trial site and recomputed again after removing one additional site. This resulted in 1,008 factors per year per group.

AADNT Estimation and Error Analysis

To calculate the estimated AADNT error for manual count sites using the factors discussed, the analysis pretended that one of the continuous counters was actually a short duration site (trial_short_duration). This made it possible to know the actual AADNT for that site and also investigate an error estimate if only 1 h that year had been counted manually. For each hour of the day (7:00 a.m. to 7:00 p.m.) for each trial short duration site, the estimated AADNT is computed using

$$\text{Estimated AADNT}_{yskhdg} = \frac{(\text{trial\_short\_duration } V_{hdks})}{(\text{trial\_group } H_{hjmykg})} \tag{4}$$

where

Estimated AADNT_{yskhdg} = the annual average daily bicyclists and pedestrians estimated for a given trial short duration site, *s*, year, *y*, mode, *k*, for the hour of the day of the trial short duration count, *h*, for trial short duration count date, *d*, and for trial factor group, *g*;

trial_short_duration V_{hdks} = volume for a particular hour, *h*, date, *d*, and mode, *k*, for the trial short duration site, *s*; and

trial_group H_{hjmykg} = hourly factor for a given hour of the day, *h*, day of the week, *j*, month of the year, *m*, year, *y*, mode, *k*, and for the trial group, *g*. The trial group, *g*, is composed of the sites in the factor group except the trial short duration site, *s*.

The trial group, *g*, does not include the trial short duration site, *s*, because the researchers do not want to estimate AADNT at a site using data from that site. The study is trying to simulate the situation in which the full year of data is not available, so including such data in the factor would reduce the error, and not be a good test of the method. These calculations are repeated for all sites.

The error was computed as the mean percentage error (MPE) and MAPE using

Table 3. Short Duration Count Scenarios

Duration of count	Time frame	Days of the week (DOW)	Source
2 h	7–9 a.m.	TWorTh	WSBPDP
2 h	3–5 p.m.	TWorTh	1-h analysis found lower error in Table 4
2 h	4–6 p.m.	TWorTh	WSBPDP
2 h	5–7 p.m.	TWorTh	NBPDP
2 h	12–2 p.m.	Saturday	NBPDP
2 h	4–6 p.m.	Any day	1-h analysis found lower error in Table 4
4 h	7–9 a.m., 4–6 p.m.	TWorTh	WSBPDP
4 h	5–7 p.m.	TWorTh	NBPDP
	12–2 p.m.	Saturday	
6 h	7–9 a.m.	TWandTh	Utah DOT
6 h	4–6 p.m.	TWandTh	Utah DOT
8 h	7–9 a.m., 4–6 p.m.	TWorTh	Utah DOT
	10–2 p.m.	Saturday	
8 h	7–9 a.m., 11 a.m.–1 p.m., 5–7 p.m.	TWorTh	Proposed new count scenario for WSBPDP
	12–2 p.m.	Saturday	
12 h	7 a.m.–7 p.m.	TWorTh	NBPDP, Utah DOT
24 h	7 a.m.–7 p.m.	TWorTh	NBPDP, Utah DOT
		Saturday	

Note: WSBPDP = Washington State Bicycle and Pedestrian Documentation Project; NBPDP = National Bicycle and Pedestrian Documentation Project; Utah DOT = Utah Department of Transportation.

$$MPE_{yskhdg} = \frac{(\text{Estimated } AADNT_{yskhdg} - AADNT_{ysk})}{AADNT_{ysk}} \quad (5)$$

$$MAPE_{yskhdg} = |MPE_{yskhdg}| \quad (6)$$

where $AADNT_{ysk}$ = the actual annual average daily bicyclists and pedestrians for year, y, site, s, and mode, k.

This error analysis calculation is repeated for each trial short duration site, s, year, y, mode, k, hour, h, and trial short duration count date, d.

Short Duration Count Scenarios

Several short duration count scenarios are examined in this study as shown in Table 3. Most of the scenarios are selected from recommended timeframes from various count programs found in the United States and Canada. All of the scenarios assumed that counts are collected during April to September. Several of the scenarios assumed weekday counts are collected on Tuesday, Wednesday, or Thursday (TWorTh) or Tuesday, Wednesday, and Thursday (TWandTh) and weekend counts are collected on Saturday, as recommended by previous research and common data collection methods.

One purpose of this study is to investigate error estimation for different hours throughout the day. Thus, many of the 2-h count scenarios are derived timeframes recommended by existing count programs or guidance to represent various hours from 7:00 a.m. to 7:00 p.m.. Several 2-h scenarios represent times when the highest counts for different travel patterns are expected (e.g., p.m. peak hours for commute patterns, midday Saturdays for recreational patterns). Two 2-h scenarios are included because the initial error analysis determined that counts in the afternoon produced the lowest error estimations regardless of other variables (e.g., day of week, hourly factor group, city).

The 4-h count scenarios were modeled from the Washington State Bicycle and Pedestrian Documentation Project (WSBPDP) and the NBPDP. Because both 4-h scenarios sample days throughout April to September, rather than specific days in late September or early October, the corresponding errors are likely to be lower than that of the actual WSBPDP or NBPDP counts.

The 6-h count scenarios are based on guidance from the Utah Department of Transportation (Utah DOT), which recommends counting on consecutive weekdays (Tuesday, Wednesday, and Thursday) during morning and evening commute peak hours.

The 8-h count scenarios are combinations of 2-h scenarios previously discussed. One 8-h scenario expands on the 4-h NBPDP scenario by adding a midday period on the weekdays from 11:00 a.m. to 1:00 p.m. and a midday period on Saturday from 12:00 noon to 2:00 p.m. This scenario allows for an estimation of AMI to be computed. Estimations of AMI can be used to determine the correct factor group, and thus calculate more accurate error estimates. This scenario is the research team's proposed new count scenario for WSBPDP.

Table 4. MAPE for 1-h Counts by Day of Week

Hour	Sun	Mon	Tues	Wed	Thurs	Fri	Sat	Avg.
7	160	134	110	111	106	109	127	122
8	103	82	72	78	73	78	86	82
9	76	82	65	66	72	69	76	72
10	74	79	66	66	66	69	94	73
11	71	74	65	66	67	68	96	72
12	67	76	79	65	71	76	69	72
13	67	76	66	73	67	73	69	70
14	69	75	63	64	63	64	68	67
15	66	66	61	61	61	60	65	63
16	66	63	59	60	62	59	68	62
17	73	61	59	65	66	67	84	68
18	95	65	73	70	69	75	147	85
Average	82	78	70	70	70	72	87	

Note: Avg. = average. Hour assumes 24-hour clock. The hours with the lowest errors are in bold.

Table 5. MAPE from 1-h Counts by Factor Group

		Hourly commute		Hourly multipurpose		Hourly noon activity		
Community	Number of sites (*n*)	MAPE (%)	*n*	MAPE (%)	*n*	MAPE (%)	*n*	Weighted average (%)
Arlington	29	48	22	69	7	NA	NA	53
Boulder	17	NA	NA	61	17	NA	NA	61
Portland	32	43	6	46	8	83	18	66
San Diego	46	108	3	59	22	76	21	70
Seattle	16	55	9	112	3	101	4	77
Mount Vernon	6	NA	NA	NA	NA	125	6	125
Total	146	NA	40	NA	57	NA	49	NA
Average (%)	NA	63	NA	69	NA	96	NA	75

Note: NA = missing data or inapplicable totals or averages.

The 12-h and 24-h count scenarios are both modeled from the NBPDP and Utah DOT guidance. Counts are assumed to be collected from 7:00 a.m. to 7:00 p.m. on Tuesday, Wednesday, or Thursday only for the 12-h scenario. The 24-h scenario counts during these same times and additionally from 7:00 a.m. to 7:00 p.m. on Saturday. It is possible to collect these counts using manual counters by dividing the time period into shifts to avoid fatigue. However, it is typical that these counts would be taken using automated, continuous counters specifically calibrated to bicycles and pedestrians.

The comparison of count scenarios calculated the MPE, MAPE, and standard deviation of the MAPE using Equations 5 and 6.

Results

Error by Hour

MAPE for AADNT estimation for 1-h counts by day of week are shown in Table 4. The findings reveal that the lowest observed errors cluster from 2:00 to 6:00 p.m. on Tuesday, Wednesday, Thursday, and Friday. The highest errors are found mostly in the morning hours from 7:00 to 9:00 a.m. throughout the week and evening hours after 5:00 p.m. for weekends. Lower error is observed on Sunday rather than Saturday, which is contrary to what other research has found. However, because it is not common practice to conduct weekday counts on Friday or weekend counts on Sunday, these days are not included in many of the count scenarios.

As shown in Table 5, the commute factor group has the lowest error for 1-h counts for all cities except San Diego. This high error in San Diego is probably caused by having only three sites in that factor group, all with error above 90%. The errors presented in Table 6 for 1-h counts by mode indicate the lowest error for bicycle-only counts.

Scenario Comparison

MAPE for AADNT estimation for each of the count scenarios by city is provided in Table 7. The 12-h and 24-hour count scenario produce the lowest MAPE for April to September with 32% and 30% error, respectively.

Table 6. MAPE from 1-h Counts by Mode

Community	Number of sites (*n*)	Bicycle-only		Bike–ped-combined		Pedestrian-only		Weighted average (%)
		MAPE (%)	*n*	MAPE (%)	*n*	MAPE (%)	*n*	
Arlington	29	54	18	NA	NA	49	11	52
Boulder	17	60	17	NA	NA	NA	NA	60
Portland	32	45	9	58	15	109	8	67
San diego	46	62	33	NA	NA	88	13	69
Seattle	16	55	9	NA	NA	106	7	77
Mount Vernon	6	NA	NA	125	6	NA	NA	125
Total	146	NA	86	NA	21	NA	39	NA
Average (%)	NA	55	NA	92	NA	88	NA	69

Note: NA = missing data or inapplicable totals or averages.

Table 7. MAPE for Short Duration Count Scenarios

Scenario	Hours	MAPE (%)						
		ARL	BOU	MVN	PDX	SAN	SEA	Weighted Avg.
12–2 p.m. Saturday	2	36	48	91	44	62	77	50
7–9 a.m. TWorTh	2	32	54	175	56	59	52	50
4–6 p.m. Any day	2	38	45	93	53	54	49	48
4–6 p.m. TWorTh	2	37	44	90	55	50	43	47
3–5 p.m. TWorTh	2	38	40	71	48	53	44	46
5–7 p.m. TWorTh	2	35	46	83	55	48	37	45
7–9 a.m. and 4–6 p.m. TWorTh	4	27	40	119	40	44	40	38
5–7 p.m. TWorTh and 12–2 p.m. Saturday	4	25	37	58	38	44	45	36
7–9 a.m. TWandTh	6	28	48	196	46	54	50	44
4–6 p.m. TWandTh	6	29	36	63	47	43	38	39
3–5 p.m. TWandTh	6	29	32	60	39	47	37	38
7–9 a.m. and 4–6 p.m. TWorTh and 10–2 p.m. Saturday	8	25	34	60	30	43	45	34
7–9 a.m., 11 a.m.–1 p.m., 5–7 p.m. TWorTh and 12–2 p.m. Saturday	8	24	33	58	29	45	43	33
7 a.m.–7 p.m. TWorTh	12	25	31	83	30	40	29	32
7 a.m.–7 p.m. TWorTh and 7 a.m.–7 p.m. Saturday	24	21	31	58	25	41	40	30
Average		30	41	93	43	49	45	

Note: Available data used in this analysis were collected from April to September. ARL = Arlington; BOU = Boulder; MVN = Mount Vernon; PDX = Portland; SAN = San Diego; SEA = Seattle.

This is probably because these scenarios average more hours of data, which result in more accurate estimates of AADNT. Both the 12-h and 24-h scenarios provide enough data to observe a daily travel pattern for a weekday (TWorTh), and the 24-h scenario expands the dataset to also include a daily travel pattern for a Saturday. Both scenarios demonstrate the lowest standard deviation of MAPE (43% and 40%, respectively) found in Table 8, indicating the least variability in error estimates. The Table 7 errors are plotted in Figure 2 to show that error usually decreases as more hours are counted, although not always as in the case of the 12-h count for Mount Vernon.

For jurisdictions without resources to conduct 12-h counts, the 8-h scenario with count periods from 7:00 to 9:00 a.m., 11:00 a.m. to 1:00 p.m., and 5:00 to 7:00 p.m. TWorTh and 12:00 noon to 2:00 p.m. Saturday results in the next lowest error of 33%. The Arlington test data demonstrate the lowest error across all scenarios, and Mount Vernon the highest.

Discussion

As others have found, estimating AADNT from less than 24 h of count data results in high error. The 2-h count error estimates are similar to Budowski's seasonal average daily bicyclist error estimates using the TMG method (32% MAPE and 44% standard deviation) for 10 bicycle counters in Winnipeg. For some purposes, such error is unacceptably high, but because little other information on bicycle and pedestrian volumes is available, practitioners may choose to use such estimates

Table 8. AADNT Estimation Error for Short Duration Count Scenarios

Scenario	Hours	AADNT Error as MPE		AADNT Error as MAPE	
		MPE (%)	SD of MPE (%)	MAPE (%)	SD of MAPE (%)
12–2 p.m. Saturday	2	14	85	50	70
7–9 a.m. TWorTh	2	20	99	50	88
4–6 p.m. Any day	2	14	86	48	73
4–6 p.m. TWorTh	2	13	85	47	72
3–5 p.m. TWorTh	2	13	78	46	64
5–7 p.m. TWorTh	2	12	84	45	71
7–9 a.m. and 4–6 p.m. TWorTh	4	15	64	38	54
5–7 p.m. TWorTh and 12–2 p.m. Saturday	4	13	60	36	50
7–9 a.m. and 4–6 p.m. TWorTh and 10–2 p.m. Saturday	8	16	53	34	43
7–9 a.m., 11 a.m.–1 p.m., 5–7 p.m. TWorTh and 12–2 p.m. Saturday	8	15	54	33	46
7 a.m.–7 p.m. TWorTh	12	14	51	32	43
7 a.m.–7 p.m. TWorTh and 7 a.m.–7 p.m. Saturday	24	16	47	30	40

Note: Available data used in this analysis were collected from April to September. SD = standard deviation.

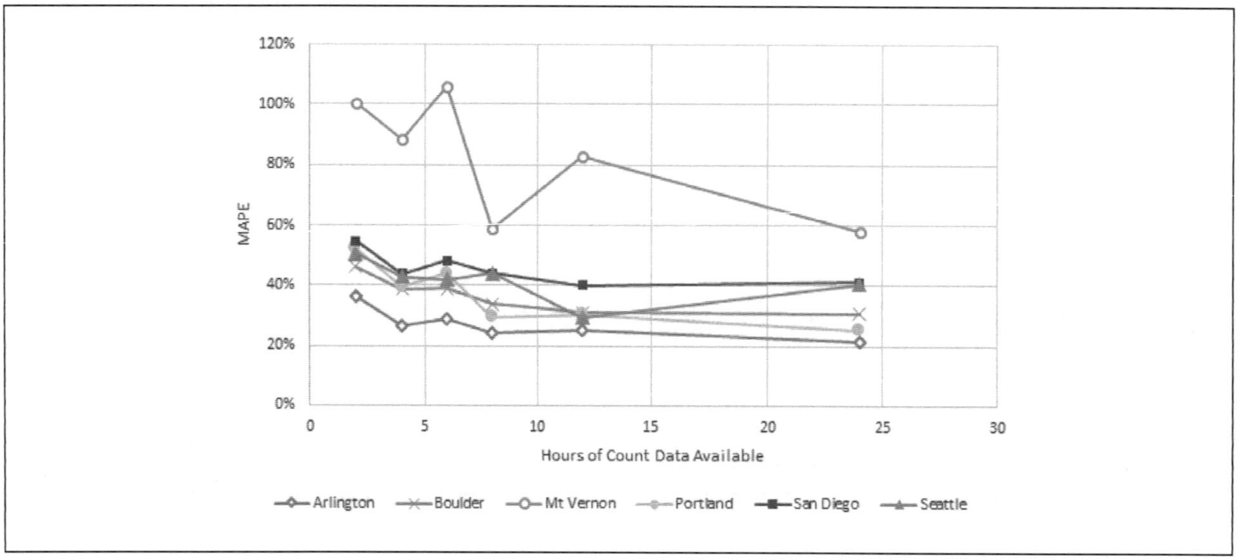

Figure 2. MAPE as a function of hours counted.

despite the error. This paper provides valuable approximations of the error for such estimates so that practitioners may make informed choices.

Error for some cities is higher than for others, as illustrated in Figure 2. For example, Mount Vernon had exceptionally high error. This is probably because of the low number of count sites (only six sites in total), the high variability of the counts (possibly because of inaccuracies in the equipment) and low volumes. However, volume does not clearly correlate with error in this dataset. Similarly, San Diego also had data quality problems, probably because of lack of equipment maintenance budget, most, but not all, of which were identified during the study's data quality process. Such problems may have caused the relatively high error. Error for Arlington is the lowest of the six cities studied, probably because of the higher number of count sites (22 in total, leading to at least seven per factor group) and the consistent data quality. It should be noted that Arlington data had been cleaned to a much greater extent than data for the other cities. This cleaning by the counter manufacturer at the request of the city included estimating data for cases in which counts were missing and removing erroneous high counts. This resulted in data being more consistent than other datasets.

Another source of error is the number of permanent counters used to create the adjustment factors. Error (MAPE) was 54% lower from estimates in which two permanent counters were used to create the factors compared with estimates in which only one permanent counter was used. This illustrates the importance of installing multiple counters per factor group. The TMG recommends at least three counters. In this study the number of counters per factor group ranged from one to 14. Future study of these data will investigate this source of error further.

Whereas Monday is associated with high error, Friday is closer to the other weekdays. This makes sense because holidays often occur on Mondays. Because error was computed for holidays, Mondays and overall error are higher than if holidays were excluded.

It is likely that because more special events occur on weekends than on weekdays, there is more variability in weekend counts, and thus higher error. However, it is unclear why Sunday produced less error than Saturday. This study was unable to determine whether more events happen on Saturday or Sunday, but this does affect the error and warrants further study.

As shown in Table 6, error for bicycle counts seems lower than for pedestrian and combined bicycle-pedestrian counts, overall, but this may be a factor of variability and data quality rather than actually associated with pedestrians. In fact, Arlington, which has relatively good data quality and relatively high volumes, shows slightly lower AADNT estimation error for pedestrians than for bicycles.

Conclusions and Recommendations

This research demonstrates that using manual counts to estimate AADNT is likely to result in errors higher than 20%. However, as this is common practice, the paper presents estimates of this error using common estimation methods applied to count data from six cities using frequently used manual count scenarios. For example, MAPE averages 45% for the often used 5:00 to 7:00 p.m. 2-h count period on Tuesday, Wednesday, or Thursday.

Error varies by number of count sites in the factor group, month, time of day, day of week, and city. It is also likely to vary by variability and volume at the count sites, and quality of the data. The commute factor group demonstrates the lowest error. Afternoon counts seem to be best for reducing error (2:00 to 6:00 p.m.). Sunday is often as good as if not better than Saturday for error, contrary to what others have found. Probably because of data quality but also nonmotorized traffic volume, Arlington had the lowest AADNT estimation error (MAPE) and Mount Vernon, Washington had the highest. Average AADNT estimation errors for the studied short duration count scenarios ranged from 30% to 50%. Error is lower for scenarios in which more peak hours are counted and when more than one permanent counter is available to estimate adjustment factors.

The following recommendations are made:

- Communities are urged to use counting equipment to count longer than 24 h to reduce the error. However, if this is not possible, the error estimates provided in this paper show how error can vary for different manual count scenarios used by jurisdictions.
- Install more than one continuous counter per factor group. Error (MAPE) decreases over 50% for factor groups with two continuous counters rather than one.
- Maintain and calibrate continuous counters. The high error from Mount Vernon illustrates the need to make sure equipment is providing high quality, valid data.
- The proposed 8-h peak hour count scenario proposed for potential future WSBPDP results in an average error of 33% (MAPE). For an 8-h count, this error is relatively low, but it is higher than could be obtained if automated equipment were used to count a week of data.

Next Steps

One aspect contributing to the error is the method used to estimate AADNT. Whereas others have used disaggregate methods to reduce error, this study employed the traditional factoring approach. It would be interesting to repeat this analysis using the disaggregate factoring method in the future.

The grouping of the sites is another aspect of the method that might affect the error. In this analysis sites were grouped only by the morning to midday ratio (AMI), but in the future using the weekend/weekday index to group sites with high weekend activity and high weekday activity would be likely to also decrease error (*13*).

When the correct morning/midday patterns and weekend/weekday patterns are known, one can group the site appropriately, but how much does the error increase when the incorrect group is chosen, or if it changes from year to year? Grouping short duration sites in the wrong factor group is another source of error that should be investigated further.

Another area of future work is examination of the number of continuous counters needed for reducing error.

Acknowledgments

The Washington State Department of Transportation and FHWA funded this work. The team at Portland State University included substantial software development work by Hui Zhang, Morgan Harvey, and Will Brown with assistance and direction from Kristin Tufte and Hau Hagedorn. The data were managed and cleaned by Portland State students led by Michelle Watkins: Carlynn De Joya and Lynn Tran. An expert panel for a parallel FHWA project provided important feedback from Jeremy Raw, Greg Lindsey, Mark Hallenbeck, and Sherry Ryan. Mike Lowry provided substantial assistance for various aspects of this project.

Author Contributions

The authors confirm contribution to the paper as follows: study conception and design: Krista Nordback; data collection: Dylan Johnstone; analysis and interpretation of results: Dylan Johnstone, Krista Nordback, Sirisha Kothuri; draft manuscript preparation: Dylan Johnstone. All authors reviewed the results and approved the final version of the manuscript.

References

1. Nordback, K., W. E. Marshall, B. N. Janson, and E. Stolz. Estimating Annual Average Daily Bicyclists: Error and Accuracy. *Transportation Research Record: Journal of the Transportation Research Board*, 2013. 2339: 90–97.
2. Alta Planning and ITE. *National Bicycle and Pedestrian Documentation Project*. http://bikepeddocumentation.org/. Accessed August 1, 2017.
3. Federal Highway Administration. *Traffic Monitoring Guide*. FHWA-PL-17-003. FHWA, U.S. Department of Transportation, 2016.
4. Griffin, G., K. Nordback, T. Gotschi, E. Stolz, and S. Kothuri. *Transportation Research Circular E-C183: Monitoring Bicyclist and Pedestrian Travel and Behavior: Current Research and Practice*. Transportation Research Board of the National Academies, Washington, D.C., 2014.
5. Ryus, P., E. Ferguson, K. M. Laustsen, R. J. Schneider, F. R. Proulx, T. Hull, and L. Miranda-Moreno. *NCHRP 797: Guidebook on Pedestrian and Bicycle Volume Data Collection*. Transportation Research Board of the National Academies, Washington, D.C., 2014.
6. Voyt, M. *Analysis of the Impact of Count Duration and Missing Data on AADT Estimates in Manitoba*. Master's Thesis. University of Manitoba, 2015.
7. El Esawey, M., C. Lim, T. Sayed, and A. I. Mosa. Development of Daily Adjustment Factors for Bicycle Traffic. *Journal of Transportation Engineering*, Vol. 139, No. 8, 2013, pp. 859–871.
8. Budowski, A. *Developing Expansion Factors to Estimate Cyclist Seasonal Average Daily Traffic in Winnipeg, MB*. Master's Thesis. University of Manitoba, 2015.
9. Figliozzi, M., P. Johnson, C. Monsere, and K. Nordback. A Methodology to Characterize Ideal Short-Term Counting Conditions and Improve AADT Estimation Accuracy Using a Regression-Based Correcting Function. *Journal of Transportation Engineering*, Vol. 140, No. 5, 2014, p. 04014014.
10. Hankey, S., G. Lindsey, and J. Marshall. Day-of-Year Scaling Factors and Design Considerations for Nonmotorized Traffic Monitoring Programs. *Transportation Research Record: Journal of the Transportation Research Board*, 2014. 2468: 64–73.
11. Nosal, T., L. F. Miranda-Moreno, and Z. Krstulic. Incorporating Weather: Comparative Analysis of Annual Average Daily Bicyclist Traffic Estimation Methods. *Transportation Research Record: Journal of the Transportation Research Board*, 2014. 2468: 100–110.
12. El Esawey, M. Estimation of Annual Average Daily Bicycle Traffic with Adjustment Factors. *Transportation Research Record: Journal of the Transportation Research Board*, 2014. 2443: 106–114.
13. Miranda-Moreno, L. F., T. Nosal, R. J. Schneider, and F. Proulx. Classification of Bicycle Traffic Patterns in Five North American Cities. *Transportation Research Record: Journal of the Transportation Research Board*, 2013. 2339: 68–79.

The Pedestrians and Cycles Section (ANF00) peer-reviewed this paper (18-05290).

Article

TRR
JOURNAL OF THE TRANSPORTATION RESEARCH BOARD

Transportation Research Record
2018, Vol. 2672(43) 145–153

DOI: 10.1177/0361198118798243
journals.sagepub.com/home/trr
SAGE

Estimating Annual Average Daily Bicycle Traffic without Permanent Counter Stations

Josh F. Roll[1] and Frank R. Proulx[2]

Abstract

Traffic volumes are a basic unit of measurement for understanding the transportation system. As investments in bicycle infrastructure are made, similar measures are necessary for understanding this non-motorized mode of travel. Methods for estimating annual average daily bicycle traffic (AADBT) are still developing, but generally employ techniques used in the motorized traffic monitoring field whereby data from permanent counters are used to construct expansion factors that are then applied to short-duration counts. This approach requires a network of permanent counters and knowledge about how to group factors into appropriate categories based on patterns observed in both the short-duration and permanent counter data. The methods presented in this paper advance a new approach to estimating AADBT solely using short-duration counts. The Seasonal Adjustment Regression Model uses statistical models that relate the daily bicycle volume to daily conditions and weather variables at a given count location. These models are then used to predict daily volumes for the remaining days of the year. To verify this approach and determine the resulting error, levels of available short-duration counts using varying amounts of permanent count data were simulated. This method was then applied to short-duration bicycle counts from Eugene, Oregon. With sufficient short-duration count data, this method can produce AADBT estimates with minimal error and without requiring a network of permanent counters. This approach also circumvents the need to determine which expansion factors should be applied to different short-term count locations by using statistical models in place of expansion factors.

Traffic volumes are a basic unit of measurement for understanding the transportation system. As increasing investments in bicycle infrastructure are made, so too grows the need to understand the impacts and operations of these facilities. Standardized traffic volume measurement uses units of annual average daily traffic, which are inferred at most locations based on short–duration traffic counts taken at the location and an understanding of the underlying temporal variation in volume based on continuous monitoring of additional locations. Methods for estimating annual average daily bicycle traffic (AADBT) typically follow a similar approach, but are still developing due to a number of challenges. For one, the underlying temporal patterns associated with non-motorized traffic are far less well understood than those associated with motor vehicle traffic. For example, although it is established that bicycle usage is generally fairly elastic with respect to precipitation, the degree to which this is the case varies considerably by geography and local conditions. Further, many communities simply do not have the resources available to invest in permanent bicycle counting infrastructure, as they are not obligated to in the same way that they must submit their motorized traffic count data to receive federal funding.

A number of previous papers have explored the effects of temporal factors, such as weather, day of the week, and university sessions, on rates of bicycling (*1*, *2*). Some of these studies consider continuous automated count data to infer travel patterns (*3*, *4*). For example, Nosal and Miranda-Moreno use autoregressive moving average models of continuous bicycle counts from multiple North American cities to show that temperature has a positive association with rates of cycling and that humidity and precipitation have negative effects, with an elevated effect due to higher rates of precipitation (*3*).

There are multiple possible approaches to the estimation of AADBT from short-duration count data. The *Traffic Monitoring Guide*'s (TMG) factoring method, for

[1]Oregon Department of Transportation, Salem, OR
[2]Toole Design Group, LLC, Portland, OR

Corresponding Author:
Address correspondence to Josh F. Roll: josh.f.roll@odot.state.or.us

instance, involves the identification of "factor groups" from a network of permanent counters, which provide insight into the typical hour-of-day and day-of-week patterns at observed count locations (*4–6*). Despite the widespread acceptance of the TMG methods for extrapolating short-duration motorized counts, alternative methods have been proposed for annualizing short-duration non-motorized counts, likely due in part to the substantial effects of weather on non-motorized traffic volumes. For example, "day-of-year" factoring methods have been proposed in which the short-duration count is scaled based on the proportion of the annual volume that the particular day(s) when the count was conducted comprises from a comparable permanent count location (*7*, *8*). Alternatively, "weather correction" models have been proposed that adjust the short-duration counts according to the bias assumed to exist in their estimates due to effects of weather, based on a bicycle volume–weather model estimated using permanent count data (*8*).

In this paper, we develop a method similar to the weather correction model described above. However, unlike that approach, the Seasonal Adjustment Regression Model (SARM) method proposed here does not necessitate any permanent count data. Instead, site-specific weather models are estimated using only short-duration counts from multiple years of data. These models are then used to interpolate the unobserved volumes for the remainder of the study years. This approach is intended for agencies that do not have the capital or political will to invest in permanent counters, but manage to incorporate short-duration bicycle counts into their regular traffic monitoring practice.

The remainder of the paper is divided into three sections. Following this introduction, the data being used are presented. Next, Section 3 describes and validates the new approach to count extrapolation using a pooled set of permanent counters throughout the Pacific Northwest, including presentation of some considerations around the number of days of data needed to achieve reasonable error and when data should be collected. Section 4 then demonstrates this new methodology using short-duration bicycle count data collected in the Central Lane Metropolitan Planning Organization (MPO) around Eugene, Oregon (OR), and compares the results with the traditional factor methodology. The final section provides a brief discussion of results, conclusions on the work presented here, and next steps for this line of inquiry.

Data Description

The two main components of this paper use two complementary data sets. Data from multiple permanent counters were used in the first section. Permanent count data was considered from six locations throughout the Pacific Northwest, including sites in Portland, Eugene, and Ashland, OR, and three sites in Seattle, Washington collected between 2013 and 2016. The Pacific Northwest is known for having relatively mild winters with consistent precipitation, and drier summers with warm temperatures. The Portland counts were collected on the Hawthorne Bridge near downtown, which is one of the main bicycle commuting bridges in the city. The Eugene data came from an off-street bicycle and pedestrian path a few miles from both the University of Oregon and downtown Eugene. Ashland's counter data came from a location near a major park outside of town. Data for Seattle was collected at a count station on the Fremont Bridge, a site north of downtown and Lake Union, and the Spokane Bridge, an east–west connector south of downtown Seattle.

Table 1 gives a summary of AADBT and the number of daily observations by year and by location at each of the permanent count sites (Table 1). None of the permanent count sites show a clear trend in volumes increasing or decreasing. Traffic volumes range over an order of magnitude, from an overall average of 155 daily bicyclists at the Ashland Dog Park site to 4,329 at the Hawthorne

Table 1. Average Annual Daily Bicycle Traffic and Sample Sizes

	Year							
	2013		2014		2015		2016	
Location	AADBT	*N*	AADBT	*N*	AADBT	*N*	AADBT	*N*
Fern Ridge West Chambers (Eugene, OR)	605	308	426	201	557	354	539	324
Fremont Bridge West Sidewalk (Seattle, WA)	1,311	364	1,414	364	1,372	363	1,283	365
Fremont Bridge East Sidewalk (Seattle, WA)	1,230	364	1,347	364	1,331	363	1,408	365
Spokane Street Bridge (Seattle, WA)	NA	NA	777	363	821	364	815	365
Hawthorne Bridge (Portland, OR)	4,670	365	4,706	364	4,620	363	3,321	346
Ashland Dog Park (Ashland, OR)	142	361	192	365	148	365	140	366

Note: AADBT = annual average daily bicycle traffic.

Table 2. Short-Duration Count Data Summary

Location	Average daily bicycle volume		Hourly R^2	WWI	Daily observations		Distance from reference (miles)
	Weekend	Weekday			Weekend	Weekday	
Fern Ridge West Chambers[a]	406	492	NA	0.825	297	715	NA
Heron Bridge South Fern Ridge	405	467	0.973	0.87	12	27	0.184
DeFazio Br. South River	391	546	0.959	0.71	8	18	1.95
DeFazio Br. North River	462	544	0.952	0.85	8	18	2.09
15th Ave. West Jefferson St.	453	621	0.946	0.73	10	24	0.857
Frohnmayer South River	714	982	0.923	0.73	10	27	2.43
Northbank South Greenway Br.	392	544	0.973	0.72	10	18	1.81

Note: WWI = weekday–weekend index; NA = not available.
[a]Reference site.

Bridge in Portland. Data gaps arose at some of these sites due to equipment issues and path closures, but there were sufficient data to work with when considering multiple years of observations. For the second component of the paper, the focus is on short-duration count sites and comparing the SARM approach with traditional factor methods described in guidance like TMG. In order to determine which locations to use in the application comparison section, short-term traffic bicycle count data from around Eugene, OR was examined to find locations where the daily patterns matched that from a permanent bicycle counter located on the Fern Ridge Trail at West Chambers Road. These particular short-duration count locations, summarized in Table 2, were selected due to having similar traffic patterns as suggested by their relatively consistent hourly patterns and weekday–weekend indices (WWIs) (Table 2). The hourly patterns were compared with the reference permanent counter based on the coefficient of determination (R^2) between the average percentages of traffic in each hour of the day. The WWI is calculated as:

$$WWI = \frac{\bar{v}_{we}}{\bar{v}_{wd}}$$

where $\bar{v}_{we}$ = average daily volume on weekends, and $\bar{v}_{wd}$ = average daily volume on weekdays.

The number of daily observations available at each site is also summarized in the table. The DeFazio Bridge locations had the fewest observed days, with 26 total days of data collected on each side of the bridge, and the Heron Bridge had the most counts available aside from the permanent counter reference site.The hourly patterns for the application and reference sites are presented in Figure 1 (Figure 1). This figure shows the similarity between the reference site (Fern Ridge West Chambers) and the application sites exhibiting a commute–recreation pattern with a peak in the morning and a wider and greater peak in the afternoon.

To provide some context for the short-duration count sites, Figure 2 shows their geographic locations throughout Eugene. As can be seen in the map, these locations are situated throughout the city mostly on off-street bicycle and pedestrian multi-use paths. The location at 15th Ave. West of Jefferson is an on-street location connector to the off-street Fern Ridge Path (Figure 2).

Seasonal Adjustment Regression Model Development

In this paper, we develop a method that uses daily conditions and observed daily counts at a set of locations to estimate annual bicycle traffic models. A separate predictive model is estimated at each count location, with the intent of controlling for the site-level characteristics affecting traffic volumes while still capturing variation in temporal variables such as weather. This approach relies on having a moderately large set of short-duration counts at each location where volumes are desired with coverage under varied environmental conditions. However, it does not require a network of permanent counters, which makes it relevant for agencies that collect data exclusively using portable counters for short-duration counts.

Model Description

To predict the daily bicycle volumes at each site, we experimented with estimating regression models with both Poisson and negative binomial error distributions. Because our focus is on accurate prediction of volumes, models were compared on their predictive accuracy as described in the validation section.

It was found that if over dispersion exists in the training data then the negative binomial specification is best at reducing error for some locations, however marginal the decrease.

In either case, the model is specified as linear-in-parameters with a log-link function:

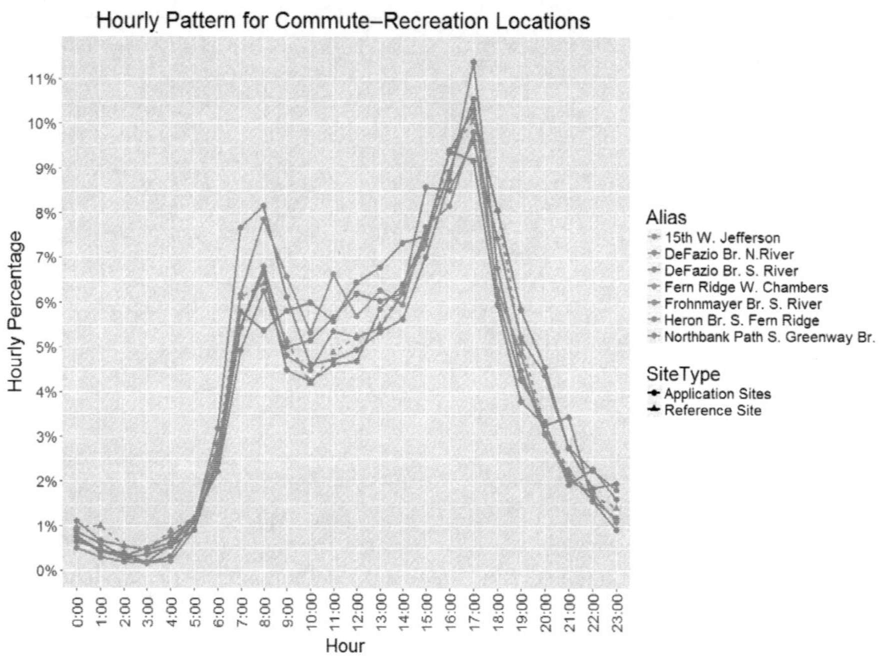

Figure 1. Hourly pattern for Fern Ridge West Chambers and traffic factor application sites.

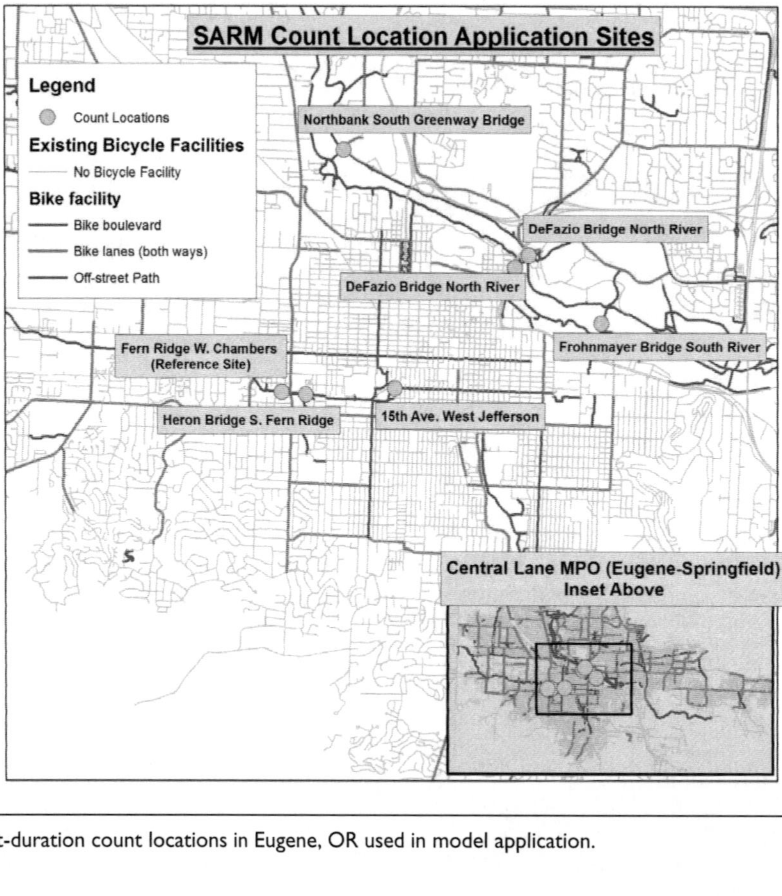

Figure 2. Map of short-duration count locations in Eugene, OR used in model application.

$$Y_{id} \sim Poisson(\mu_{id}) \text{ or } Y_{id} \sim NegBinom(\mu_{id})$$
$$\log(\mu_{id}) = \beta_i X_{id}$$

where

Y_{id} = 24-h bicycle traffic volume on day d at site i;

β_i = vector of parameters for count site i; and

X_{id} = vector of observed covariates for count site i on day d.

The specific variables included in X_{id} are the same for all count sites, and include the maximum daily temperature, total daily precipitation, minutes of daylight, an indicator for whether or not the local university is in session, whether the day is a weekday or weekend, and a factor variable for the year. If for any site a given variable lacks sufficient variation on the observed days to estimate a model, that term is omitted from the model specification. For example, a site might not have any observations with precipitation, in which case the model would be estimated without considering precipitation.

Validation

The accuracy of the proposed approach is tested by using a set of Monte Carlo cross-validation tests. Monte Carlo cross-validation involves randomly selecting a small subset of the data, estimating a model on that subset, comparing predictive accuracy against a wholly independent AADBT from a different year and repeating a specified number of times with replacement. We performed 500 iterations for our validation testing, and experimented with varying the number of weeks and years used in the simulated "data collection" period to reveal the effects of data collections strategy on predictive accuracy for this modeling approach.

The validation process works in three steps. In Step 1, short-term count data (between 2 and 6 weeks) are randomly selected from the full data set. In Step 2 a regression model is estimated and the parameters applied to a calendar of 365 days of weather and other daily conditions. Finally, in Step 3 the absolute percent error (APE) is calculated by comparing the estimated annual bicycle volume with the observed (from the independent data set).

The following equation details the APE calculation. The above process is repeated 500 times for a given count location to assess the variability in the APE:

$$APE = \left| \frac{AADBT_{obs} - AADBT_{est}}{AADBT_{obs}} \right|$$

where $AADBT_{obs}$ = observed AADBT for the count location, and $AADBT_{est}$ = Model-predicted AADBT for the count location.

Constraints were placed on the random selection of training data. First, we verified that there was at least one daily observation on both weekdays and weekends and second, we ensured that a specific set of test data was not used multiple times. Finally, test samples were drawn as contiguous weeks of data to simulate the way that data would actually be collected in the field.

Guidance on traffic counts generally directs technicians to collect count data in spring and summer, and fall but these validation tests use all months with no pretenses about which parts of the year may be best for collecting data. This approach to testing is conducted using the permanent count data, as this data set allows us to compare predicted AADBT with observed AADBT.

Results

The results from this study are described in two separate parts. First, we apply the proposed modeling approach to the permanent count stations described above. We explore the characteristics of the model's predictive accuracy in various ways for these locations. Next, we apply the approach to the short-duration counts in Eugene, OR and compare the predicted AADBT values against values estimated using the TMG's factoring approach.

Model Verification with Continuous Counts

The graphic featured in Figure 3 shows the median APE for 500 iterations of tests using different scenarios to test the seasonal adjustment methodology. Scenario differences include varying the number of weeks and years used in the training data, which varies the number of days. For example, in the scenario with 3 years and 2 weeks, 42 days of data would be used, or 14 days from three different years. Figure 2 demonstrates that median APE for all the tests done in each scenario improves significantly with 3 years of data compared with 2 years of data.

Using fewer weeks of counts from more years generally reduced the error for sites used in this examination. Error is not uniform across sites however, with the most inaccurate estimates resulting in 67% APE at the Ashland Dog Park location and the lowest amount of error of 2% at the Fremont Bridge West Sidewalk, and Spokane Street Bridge location. The Hawthorne location presents a case whereby adding more days to the training data results in error increases. This is likely due to the substantial drop in traffic in late 2015 with the opening of the Tilikum Bridge in Portland, which allowed for some bicycle traffic to divert to the new facility. As can be observed in Table 1 above, the AADBT drops significantly in 2016 compared with previous years. As the validation process uses data from 2013, 2014, and 2016 it is likely that as more weeks of data were used in the training data, more of the days came from 2016, reducing the ability to estimate the 2015 year. This is perhaps a

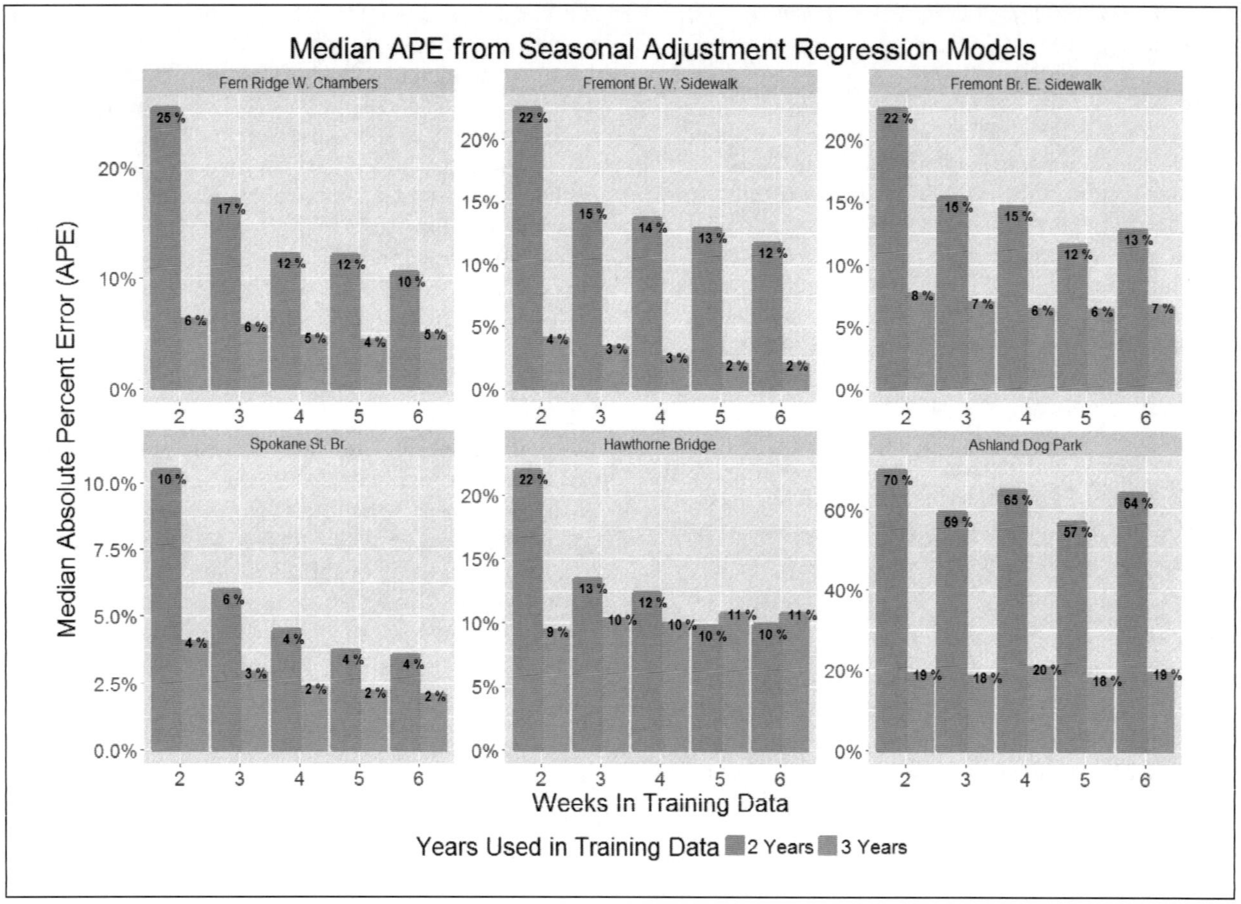

Figure 3. Median APE from multiple seasonal adjustment regression models (negative binomial).

Table 3. Deciles of Error in Seasonal Adjustment Regression Models

Location	10%	20%	30%	40%	50%	60%	70%	80%	90%
Fern Ridge West Chambers	1.25%	2.5%	3.7%	4.9%	6.2%	7.9%	9.8%	12.4%	16.6%
Fremont Bridge West Sidewalk	0.73%	1.6%	2.6%	3.6%	4.5%	5.6%	7.1%	8.4%	11.8%
Fremont Bridge East Sidewalk	1.27%	2.8%	4.2%	5.7%	7.0%	8.7%	10.6%	12.6%	15.5%
Spokane Street Bridge	0.79%	1.5%	2.4%	3.4%	4.5%	5.6%	6.9%	8.5%	12.0%
Hawthorne Bridge	3.37%	5.3%	6.7%	8.1%	9.3%	10.5%	12.3%	15.4%	20.8%
Ashland Dog Park	2.55%	5.5%	8.9%	12.4%	16.5%	21.7%	27.2%	36.3%	57.4%

cautionary note on applying these techniques when traffic levels are known to have changed substantially (Figure 3).

The lowest error estimates derive from using 6 weeks of data from three sample years, or 126 days of data, which is probably an unlikely condition for any short-term count program. Therefore comparison of the 2 week, 3 year scenario will compose the remainder of the validation test result discussion.

Table 3 shows errors by decile for all the count sites to demonstrate the range of error from the multiple (500) iterations of holdout validation tests. Each decile represents the amount of APE associated with that proportion of tests. For example, for the Fern Ridge West Chambers, 50% (median) of the tests results had 6.3% APE or less. These results are also cumulative so each subsequent decile includes the one before it. Looking at the validation tests this way, readers can see that for some locations 90% of the tests result in error of 12% or less as is the case for the Fremont Bridge West Sidewalk location, but as much as 57.4% or less in 90% of the tests as is the case for the Ashland Dog Park location.

Table 4. SARM Results for Selected Months

Months used in SARM	Spread of data collection	Fremont Bridge West Sidewalk	Fremont Bridge East Sidewalk	Spokane Street Bridge	Hawthorne Bridge	Ashland Dog Park
All months (Jan–Dec)	Low spread	29%	29%	31%	20%	56%
	High spread	18%	19%	17%	16%	55%
	Median	20%	21%	20%	17%	56%
NBPD months	Low spread	NA	NA	NA	NA	NA
(May, September, October)	High spread	22%	27%	22%	6%	79%
	Median	22%	27%	22%	6%	79%
Nordback et al. (9)	Low spread	24%	34%	23%	14%	49%
(July, August, September, October)	High spread	NA	NA	NA	NA	NA
	Median	24%	34%	23%	14%	49%
March, April, May,	Low spread	15%	16%	19%	8%	92%
September, October	High spread	14%	17%	14%	8%	68%
	Median	15%	16%	17%	8%	70%

Note: NBPD = National Bicycle and Pedestrian Documentation Project; NA = not available.

The distribution of the error described in Table 3 is a product of the months used in the SARM and shows that error can vary depending on which data are employed to estimate the model.

Table 4 summarizes another set of validation tests in which different combinations of months are used to estimate the SARM with associated APE presented. The table compares results using months that previous guidance has recommended with two other sets of months including a non-restrictive scenario in which all months are available and one that results in some of the lowest error results. Previous guidance form the National Bicycle and Pedestrian Documentation Project (NBPD) suggests bicycle counts be collected in May, September, and October so the application of SARM using these months is presented. Additionally, Nordback et al. concluded that using the months of July, August, September, and October are best to reduce error when using traditional traffic factoring methods (*9*). Presenting results from SARM validation tests in this way should be helpful to understand the expected accuracy of SARM results based on when short-term data collection would take place (Table 4).

In addition to testing different months the table breaks down the results based on proximity of the months to one another, or what is described as spread. This attribute is a computed quantity in which the time (distance) between months is measured and averaged resulting in a value between 1 (lowest spread) and 4 (highest spread). For instance, if data was collected from January, May, and August, the spread value would be greater than the if the months April, May, and June were used to calculate the spread value since the latter months are all consecutive months.

The results presented in Table 4 demonstrate which months used in the SARM would produce the lowest error. It should be noted that the results presented here simulate using 2 weeks of data collected in three different years to predict the AADBT. Therefore it could be expected that if 6 weeks of data were collected across 3 years with little variation in the months' uses (e.g., June, July, August) the SARM approach could predict AADBT for the Fremont Bridge West Sidewalk location within 29% APE, however if short-term counts were gathered aiming to spread out the collection (e.g., January, May, August), expected AADBT error from a SARM estimate would be reduced to 18%. In nearly every case, error is reduced by using a higher spread of data in the SARM model. In most cases the SARM approach using TMG months does produce better results compared with tests using all the months. This is not true for the Hawthorne Bridge location in which the TMG months produce the best results.

For most locations, using the months March, April, May, October, and September produces the least amount of error and, in most cases, the tests using these months but spread out the data collection achieve even lower amounts of error. The Ashland location is one exception to these results: the tests that use all the months of the year produce the least amount of error. Another exception to these months being best for reducing error is the Hawthorne Bridge location for which the NBPD months are best at reducing error.

Comparing SARM with TMG Factoring Method

Validation of the SARM method above uses permanent count data and a set of Monte Carlo hold out tests to assess the accuracy of the approach by comparing with observed AADBT from an independent year of counts. The following section will employ the SARM method using a select set of actual short-term count data and compare with an

Table 5. Comparison of Traditional Expansion Factor and SARM Estimates of AADBT

Location	AADBT estimate		APE	Days used in estimate		Monthly factor information	
	Traditional	SARM		Traditional	SARM	Factor	Month
Heron Bridge South Fern Ridge	398	415	4.3%	7	45	0.984	April
DeFazio Br. South River	484	457	5.6%	7	33	0.83	September
DeFazio Br. North River	445	453	1.9%	7	25	0.83	September
15th Ave. West Jefferson St.	556	513	7.8%	7	34	1.29	February
Frohnmayer South River	749	741	1.1%	7	47	0.835	May
North bank South Greenway Br.	464	471	1.6%	8	43	1.479	November

Note: AADBT = annual average daily bicycle traffic; SARM = seasonal adjustment regression model; APE = absolute percent error.

estimate derived using the traditional traffic factoring approach described in the following equation:

$$AADBT_e = ckh*Dyf*Myf$$

where

AADBTe = estimated annual average daily bicyclists;

ckh = known count for 1 h;

Dyf = daily factor for given day of week in given year y for factor group f: (actual AADBT for that year)/(average daily bicyclists for that day of the week in that year); and

Myf = monthly factor for given month in given year y for factor group f: (actual AADBT for that year)/(average daily bicyclists for that month of that year).

In theory the SARM method can be applied to any location provided sufficient numbers of days of data are available. However, guidance on application of traditional factors suggests matching hourly patterns and weekend to weekday indices (*9*). Using a single permanent counter located in the Central Lane MPO on the Fern Ridge Path, short-term sites were selected based on these two criteria.

As the hourly distribution pattern and the WWI are similar, it is reasonable to apply a monthly factor created using the Fern Ridge West Chambers site to the application sites in order to compare with AADBT estimates from the SARM approach. Table 5 shows the results of this comparison along with the APE between the estimates and other summary information including the number of daily observations used in both methods and the monthly expansion factor used in the traditional factoring process (Table 5).

According to the results presented in Table 5, the two AADBT procedures produce similar results with differences as small as 1.1% and as great as 7.8%. The two approaches have different data requirements with the traditional approach using as little as seven daily observations, whereas the SARM process uses up to 47 days of data. The comparability of these AADBT results should lend confidence to the SARM approach.

Conclusions

In this paper, we have developed, validated, and demonstrated a novel approach for estimating AADBT at a set of count sites that does not require data from a permanent counter. Our approach entails developing a daily count prediction model for each count site based on short-duration counts taken at the site. This approach requires collecting moderately long counts at each site, but this approach may prove more affordable or politically feasible for agencies that do not have the capital available to install multiple permanent count stations, as longer counts simply require leaving short-duration counters in the field for longer time periods. In lieu of permanent counters, agencies may instead spend resources on staff to move their portable counters in order to gather data from more locations, helping to understand bicycle traffic patterns at a wider geographic scale.

This research has implications for when agencies may want to be collecting data using portable counters. The SARM results presented in Table 4 show that data collection does not need to be restricted to certain months and, depending on the desired level of accuracy of AADBT estimates, data could be collected throughout the year as long as data were collected in a variety of months.

This approach is certainly more data intensive, requiring multiple years of data in order to be most effective. This research examined multiple scenarios but the most feasible is probably for agencies to collect bicycle counts for 2 weeks a year over 3 years. This approach would not be impossible and for agencies without plans for permanent counters, could be a reliable path forward that may help to balance the cost of permanent counter installation with the need to gather data in more geographically diverse locations.

The statistical techniques applied above could seem complicated to some agency staff, however tools could be developed to aid the development and application of the models. Certain traffic data database and visualization software already do some of the traditional factor development, therefore application SARM methods could be built into even less sophisticated tools in Excel or RShiny.

Installed permanent count stations can cost between US$8,000 and US$15,000 with ongoing battery and automated data collection costs, whereas portable bicycle counters can be purchased for less but require much more staff time to move and manage the data. Future research should continue to examine the tradeoff between AADBT estimation accuracy and spatial representation, and cost of data collection.

Future research could test the application of the SARM approach with traditional factoring techniques using permanent count data so the observed AADBT could be compared to determine the actual accuracy. This work only compared the results from the application of both approaches to one another, and not with the observed values.

Acknowledgments

This work was funded by the Oregon Department of Transportation and data was collected by staff and the Lane Council of Governments.

Author Contributions

The authors confirm contribution to the paper as follows: study conception and design: JR, FP; data collection: JR; analysis: JR; interpretation of results: JR, FP; draft manuscript preparation: JR, FP. All authors reviewed the results and approved the final version of the manuscript.

References

1. Tin, S. T., A. Woodward, E. Robinson, and S. Ameratunga. Temporal, Seasonal and Weather Effects on Cycle Volume: An Ecological Study. *Environmental Health*, Vol. 11, No. 2, 2012, p. 12. https://doi.org/http://dx.doi.org/10.1186/1476-069X-11-12.
2. Rose, G., F. Ahmed, M. Figliozzi, and C. Jakob. Quantifying and Comparing the Effects of Weather on Bicycle Demand in Melbourne (Australia) and Portland (USA). Presented at 90th Annual Meeting of the Transportation Research Board, Washington, D.C., 2011.
3. Nosal, T., and L. F. Miranda-Moreno. The Effect of Weather on the Use of North American Bicycle Facilities:A Multi-City Analysis using Automatic Counts. *Transportation Research Part A: Policy and Practice*, Vol. 66, 2014, pp. 213–225. https://doi.org/doi:10.1016/j.tra.2014.04.012.
4. Miranda-Moreno, L., T. Nosal, R. Schneider, and F. Proulx. Classification of Bicycle Traffic Patterns in Five North American Cities. *Transportation Research Record: Journal of the Transportation Research Board*, 2013. 2339: 68–79.
5. Federal Highway Administration. *Traffic Monitoring Guide*. U.S. Department of Transportation, Washington, D.C., 2013. https://www.fhwa.dot.gov/policyinformation/tmguide/.
6. Roll, J. F. Bicycle Traffic Count Factoring: An Examination of National, State and Locally Derived Daily Extrapolation Factors, 2013. https://pdxscholar.library.pdx.edu/open_access_etds/998/.
7. Hankey, S., G. Lindsey, and J. Marshall. Day-of-Year Scaling Factors and Design Considerations for Nonmotorized Traffic Monitoring Programs. *Transportation Research Record: Journal of the Transportation Research Board*, 2014. 2468: 64–73.
8. Nosal, T. G., L. F. Miranda-Moreno, and Z. Krstulic. Incorporating Weather: A Comparative Analysis of Average Annual Daily Bicyclist Estimation Methods. *Transportation Research Record: Journal of the Transportation Research Board*, 2014. 2468: 100–110.
9. Nordback, K., W. E. Marshall, B. N. Janson, and E. Stolz. Estimating Annual Average Daily Bicyclists: Error and Accuracy. *Transportation Research Record: Journal of the Transportation Research Board*, 2013. 2339: 90–97.

The Pedestrians and Cycles Section (ANF00) peer-reviewed this paper (18-05673).

Sage Publications USA
sagepub.com

ISBN 978-1-5108-8249-2

9 781510 882492

CURRAN ASSOCIATES INC.
proceedings.com

2nd Workshop on Natural Language Processing and Linked Open Data 2015 (NLP&LOD2)

International Conference on Recent Advances in Natural Language Processing 2015

Hissar, Bulgaria
11 September 2015